PRACTICAL INVENTORS GUIDE:

SUCCESSFULLY NAVIGATING THE COMPLEX PATHWAY

FROM GREAT IDEA TO PRODUCT FOR SALE

Stephen M. Samuel PE and Ashlyn Wright

ISBN: 978-1-935951-08-7

DESIGN VISIONARIES
TEL: 408.997.6323

EMAIL: INFO@DESIGNVIZ.COM

WWW.DESIGNVISIONARIES.COM

DEDICATION

To Julia, who inspires me to commit every fiber of my being to be the best dad I can and who motivates me to try as hard as I can to help bring peace and prosperity to as many people as I can touch. Thank you love. You are an amazing person.

TABLE OF CONTENTS

PREFACE

This handbook compiles decades of experience working with some of the most creative people in the world to get their products to market. As a team, we at Design Visionaries have many credits to our name and have developed some truly wonderful and exciting designs for products on the market today. From incredibly high-tech aerospace components to the tips of earbuds, we've done it all. We hope to bring much of what we've learned to you so that your journey from idea to market is as smooth and low cost as possible.

Practical Inventors Guide is also the culmination of countless hours of conversations we've had with many of the most sought-after advisors: electrical engineers, model makers, sheet metal producers, patent attorneys, and beyond. The questions that we asked all have to do with the same general theme: What does our creative community need to know in order to get their product on the market in the most efficient way? We set up these conversations to increase our own knowledge and better our designs; now we're sharing what we've learned with you. We've got tips on every aspect of the invention process, from generating good ideas to showing off your product to buyers, so pick the section most useful to you and get reading!

INTRODUCTION: THE WISDOM OF MENTORS

"As a mechanical engineer, you are an arranger," he'd say. "You may be arranging for 300-degree air to be delivered to a certain place at a certain rate, or you may be arranging for five gallons per minute of cooling water to be deposited somewhere, but as an engineer you have to take care of all the details, all the components have to work together to get a task or series of tasks done." That was the inspiring lesson I learned from one of my favorite engineering professors, Professor Heronomous. He was a large man, and he exuded a combination of strength and warmth that I can only describe as fatherly. I understood that he had been a military man of some sort, perhaps in the navy. At one point he told the class that his favorite color was battleship gray. I was incredibly fortunate to have had him as a senior design class instructor. I looked up to him, and he gave all of us college kids his wisdom.

What he told us was true. We really were learning to be arrangers. Mechanical engineering in many cases *is* the arranging of all sorts of things. Almost every great mechanically engineered product is the result of a number of pieces that are in their own way the culmination of thousands of years of human technology. As soon as you take two sheets of anything and screw them together, you are using the fruits of the labor of all the people who worked before you to figure out the best angles for the threads, the material science, and the way to make a machine that created those screws at a rate of a kajillion per minute. It can be easy to forget this with how easily screws are bought off the shelf now, but that was just one important lesson I learned from one of the many great mentors in my life.

Professor Boothroyd was a tough mentor. He was an ex-military man from England who seemed to have a bit of disdain for the happy-go-lucky students that would show up to class in sandals. I had him for a class called Statics. On the face of it, it looks like a very easy class. Most of what you learn is how to keep things still. How many times did we hear, "The sum of all the forces equals zero, the sum of all the moments equals zero."? However, many of the students had a very hard time with the class. It was being used as a weed-out class.

See, the University of Massachusetts in the '80s was a great place. It was also known as a party school. It appeared that they followed the philosophy of letting many people into the engineering program, then they made the program in some ways harder than it had to be in order to filter out a much smaller number. In freshman orientation one of the professors gave a classic statement: "Look to the left of you. Look to the right of you. One of you is not going to be here in the next few semesters." The year I took the Statics class from Professor Boothroyd, a full 50 percent of the students failed. I was one of the fortunate ones who passed.

The great thing about Professor Boothroyd was his ability to show students how to focus on what was important. In Statics it's very important to look closely at the configurations of the various structures you are analyzing and try to see in advance what portions of that structure have nothing to do with what you're trying to find out. For example, if you're trying to figure out how large to make an engine mount, there are ways to lump all the complex components together into one "point mass" so your calculations will be easier. Professor Boothroyd also took great care to teach us to look at how things were connected and whether they were pin jointed or cantilevered to see what members could be ignored. "When you're trying to figure out how much force is on each member of a truss bridge, first figure out which ones have no bearing on what you are trying to find," he'd remind us. That lesson has meaning in a broader sense as well: Figure out what's important, and give it all your focus.

One fine night we had a Statics midterm scheduled for 6:00 P.M. I studied until about 4:00 when I decided to take a short nap. My alarm clock didn't go off and I ended up waking at 6:15. I panicked. I threw my clothes on, grabbed my knapsack, and ran all the way down to where the test was being held. By the time I arrived it was already 6:25. In my haste I forgot my calculator. The first professor I encountered was Boothroyd. I threw myself at his mercy and told him that I had woken up late and forgotten my calculator. I pleaded with him to let me take a make-up exam. He seemed to smile at me but I soon discovered that it was actually a mean sneer as he said, "Young man, just because you're late that doesn't excuse you. You have to take the test anyway. And if you don't have a calculator"—he held his fingers up to my face——"just use these." Without missing a beat, I walked over to Professor Heronomous and told him that I didn't have a calculator. He lent me his and told me not to break it.

In the end I did very well on the test that 40 percent of the students failed. As Boothroyd handed back the exams some days later he smiled at me. I think he was genuinely happy that I did well. I think he was trying to teach me to be strong and responsible just as a drill sergeant might give his privates a real hard time to toughen them up. The reduced amount of time that I had to take the exam didn't hurt me. It forced me to use the skills that Boothroyd had taught. I learned his lessons well; I knew what to focus on and to eliminate all the things that really didn't matter. Thank you, Professor Boothroyd, wherever you are.

Professor Umholtz had at least one product on the market that I knew of and had a patent on a process for forming metal with explosives. I had him for a course on engineering drawings. He was artistic and creative and a very pleasant guy. He was all about the feeling. Watching him work with a piece of chalk on the board was a true learning experience. His whole body seemed to be involved as he drew beautiful shapes. The big lesson I learned from him was focus. When he drew it was as if nothing else existed in the universe. The lesson was clear: A good design is facilitated when one can focus as if there was nothing else on Earth except the one task at hand.

My list of mentors goes beyond just college professors. As far as design engineering is concerned, the greatest mentor I ever had was my father David F. Samuel. He was an amazingly creative man. He had grown up in a large family in the '20s and '30s when money was scarce. He and his siblings would make their toys. They would also find toys that others discarded, fix them, and use them as their own. My father could sketch, paint, carve, calligraph, and do just about anything else he put his mind to. He built a sailboat, dulcimers, and puppets out of papier mache and myriad other things. He was also a captain in the army reserves so he had good discipline, too.

My father was from a generation that didn't trust art as a way to make money for a family, so he became a teacher and eventually a principal, but when he retired he decided to become a commercial artist and achieved a respectable level of success. He had some of his painted works in museums in Florida and Baltimore, and I was intensely proud of him both when he was living and now. His main message was this, "Study a thing—do a thing." Basically he believed that there was nothing you couldn't do if you just got the right instruction and committed yourself to getting it done.

When I was about 11 he purchased a five-acre piece of land in New York State with a shack on it. The shack had electricity, but no running water. It had a well with an old-fashioned hand pump and an outhouse. Before long, we had purchased a stack of home building books and built on a huge addition, a Jacuzzi room, and a free-standing workshop. We had to hire someone to put in a septic system and dig a deep well, but other than that we did everything ourselves with what we learned from books or figured out on our own. Trips to Montgomery Ward, a Sears rival, were like going to a museum, a church, and a candy store all in one for my dad, mom, brother, and me. We would pick out plumbing components and building supplies of all kinds. We'd load large sheets of plywood on the top of a homemade roof rack that we tied to the top of our Cadillac.

Dad would ask me to do things without telling me how to do it. He would watch me figure it out and wouldn't necessarily say anything when I got it wrong at first. He would give me praise for trying, and he always looked on with quiet pride. He was the powerful voice of reason for me as I navigated through a slew of wood projects, art projects, and life in general. His overwhelming message to us boys was, "You guys can do that." We once saw a news article about a priest who had encouraged a group of kids to learn to ride unicycles. "You guys can do that," my father said. Both my brother and I can now ride a unicycle.

I knew that I was supposed to be good at making things. I may have become good due to that expectation or it may have been innate or both, but from an early age I put pressure on myself to be able to envision and build things. One of the first big projects I can remember building was a chair. I must have been about six years old. I had an old saw, a hammer, nails, and wood from someplace. I sawed and hammered and put this thing together for hours. Where does a six year old get the patience to do something like that without supervision? What six year old these days gets to be unsupervised and gets to think of themselves as a person who can answer their own questions?

Another extremely powerful mentor in my life has been, and is, my older brother Mike. My brother and I grew up in a time when TV wasn't very good. There were only a few channels and a few cartoons and long periods of time where there was nothing good on to watch at all. So Mike and I would spend hours drawing and playing games with other kids instead. We both became successful athletes in our own right and were both considered very creative. Mike is, and always was, a naturally talented artist. He could do anything. At an early age, his cartoons were professional quality. If anyone needed a cartoon or a poster or even something cool drawn on their sneakers, they wanted Mike to do it. Almost everyone reading this book has seen my brother's work. If you've seen the H logo for the History Channel, or the Sears logo, or other logos for many well-known products you've seen the result of Mike's natural talent and his will to practice his skill incessantly. He is never without a pen and some paper. I heard him say early in life, "Talent is the will to practice."

I was so amazed when I started work in the early '80s at the Pratt and Whitney jet engine manufacturing plant in East Hartford, Connecticut. Pratt and Whitney, at the time, was like a college for me and the new influx of college graduates hired in 1983. There was a group of us, young and single, that used to go to happy hour on Friday afternoons. I was extremely fortunate to land the job and even more fortunate that the field of computer aided design was just taking off. At that time CAD (Computer Aided Design) systems were ultra-expensive. I was on the team at Pratt that was instrumental in finding and evaluating new CAD systems and helping to get them accepted in the general population. At an early age I was given the rare opportunity to learn these systems and work on the ground floor, so to speak. Pratt and Whitney and the opportunity I was lucky enough to gain was an amazing experience that had a big influence on me. It pushed me forward just as all those other great mentors did.

Later on, I parlayed my skills and knowledge into a successful design consulting business of my own and have been running it for 26 fun- and work-filled years. This book is a product of those years of experience, which would not have been possible without the support and lessons I learned from those great influences in my life. I hope you get the chance to have your own great mentors throughout your journey as an inventor. Learn from them, absorb their wisdom, and carry those lessons with you as you move through your life.

—Stephen M. Samuel

THE ANATOMY OF A GOOD IDEA

Chances are, if you're reading this book, you already have a great creative idea in mind. If so, good for you! Hopefully, as the creative visionary you are, you aren't content to settle with having just one creative epiphany in your lifetime or even in the course of this one idea. Success in inventing often comes through incredible amounts of trial and error, scrapping and rebuilding, and failures that send you right back to the drawing board. The whole process of inventing, from the idea's inception to its final form requires creativity in problem solving, design, and everything in between. It requires the persistence to push through all that to reach your ultimate goal: a successful product. But how do you know if you have a good idea, one worth investing your time, money, and energy in? A good idea is a creative one that, with the right work and circumstances, will succeed when it graduates from the realm of imagination into reality. There are a few things to look for that help ensure this is the case.

WHAT MAKES A GOOD IDEA?

As you surely know, creativity is central to engineering, design, and invention as a whole, but creativity itself is a slippery thing to define and harness. To make sense of the subject, let's look at a common definition used by creativity researchers. It's twofold, classifying creativity as something requiring both novelty and usefulness. Another definition for creativity proposes three aspects, adding surprise, or non-obviousness, as a third quality.

Novel, useful, and non-obvious. If you've looked into patenting recently, you might recognize those words. They were taken directly from the criteria used by the US Patent Office to judge whether to award a patent for an idea. As you develop your ideas or just try to figure out if they're any good, it's important to keep these three criteria in mind.

Novelty is the most obvious part of the equation. If the idea has been done before, it's not creative, just repetitive. Even novelty relies on a certain amount of convergence. Amazing ideas can be created by building off of both the old and the new. There are very few existing products that can't be improved in some new way. Something novel can be born from a new perspective or simply by combining disparate parts in a new way. Perhaps there's a new technology that, when combined with an old idea, makes a huge opportunity for a new product. Maybe there's a new material or a new behavioral pattern to work with. The ever-evolving state of the world around us brings new opportunities for novelty every second, we just have to find them.

Novelty on its own is not enough though. Sometimes there is a good reason why something hasn't been done before. A cat walking across a keyboard might make something spectacularly novel yet incredibly useless. A novel idea is nothing without some aspect that makes it useful in the current market to fulfill a current need.

They say Archimedes invented the basics of calculus over 1500 years before it was reinvented by Sir Isaac Newton and Gottfried Wilhelm Leibniz. At the time, few if any scholars continued his work on the subject, and it was lost to the world until it was recovered in 1910 so that we modern folks could examine it and gain insight into his thinking. It's fun to imagine how life would be different had humankind paid more attention to Archimedes. Many of us already know about the inventions that to this day bear his name, such as the Archimedean screw and the Archimedean principle of buoyancy. Had we as a species understood the principles of calculus 1500 years earlier, I think things would be very different today. Not necessarily better, but certainly different. Archimedes' ideas were too far ahead of his time and there weren't enough people around who could understand how amazing they were. His contemporaries failed to see their utility, so when Archimedes was killed, much of his knowledge was lost for centuries.

The utility of an idea can be boiled down to a simple statement: Good design is something that fills a need. A good idea or innovation is something that brings value to people and enhances users' lives, bringing them happiness or a feeling of wellbeing. In some cases it gives them more life. A good idea doesn't have to be earth shattering. It could simply bring some modest amount of benefit to a reasonable number of people and do just fine. It could be a jewelry box with compartments that rotate and also automatically clean their contents. It only serves the number of people who have so much jewelry that they need a box for it and who care enough about it being clean. A truly great idea, however, comes out of a great need and serves an innumerable number of people in a way that is truly life changing. Consider, for example, what the invention of seatbelts and disc brakes did for automobile safety.

Timing, especially in this day and age, is an important part of utility. As with Archimedes, time and again you'll hear stories of inventors, artists, and writers who were a bit too "ahead of their time." Their ideas may have been genius, but they gained little success in their lifetimes because others could not see the value of their creation. The pace of today's world can be helpful in this respect. Usually when there's a great need it's because of some sort of recent change, and today's world is always changing. Before there was the automobile, there was no need for seatbelts. Now, as many folks begin to accept self-driving cars, there may be an entirely new industry built up around all the things that people are going to do with the extra time that they have when they are not holding onto the wheel and looking out the window. Fads and fashions are changing all the time, often in baffling ways. If everyone suddenly wanted to make sure their tattoos could be seen at night, there might be a new market for wearable and long-lasting personal lighting. Maybe you could invent a way to implant extremely low energy LEDs under the skin. Sounds nuts? Just wait and see. New needs and new opportunities pop up constantly, just waiting to be seized.

Non-obviousness is the third bit to consider when making a good (and patentable) creative idea. A new idea that follows directly from something that already exists, something that requires no leaps of insight or ingenuity, is neither creative nor patentable. The US Patent Office defines this quality as being "Non-obvious to a person having ordinary skill in the area of technology related to the invention." It can't be a "trivial" alteration. For example, color or size variations aren't typically patentable. A word of caution for when you're trying to figure out if your idea is obvious or not: Everything looks obvious once you already know the answer! This "hindsight bias" is something both inventors and those who approve their patents need to be well aware of.

Finally, when considering what makes a good idea, it is important to be realistic, of course, but don't kill your potential with your own naysaying. Never, never, NEVER think "If this idea were that good someone would have thought of it by now." While it is healthy to be self-critical and important not to kid yourself, you must acknowledge that there are always new products and services needed by people whose lives change rapidly. Remember what we said about timing being important for utility? Your idea may take advantage of that perfect opportunity at the right time for it to be successful. Have confidence in your idea and your creative ability!

CREATIVE MOTIVATIONS

Creativity can often be seen as an outlet for emotional expression, especially when we are in situations of stress and trouble. The music business is replete with folks in pain who create these amazing works of great beauty. In this sense, the myth of the starving artist may have a basis in truth. How many times do we hear about great artists driving themselves to terrible exhaustion, dealing with serious addictions, or even committing suicide? It is a connection often given voice by artists themselves, and many think of their suffering as essential fuel for their art. Edvard Munch, the artist most famous for his painting entitled *The Scream*, once wrote:

"My fear of life is necessary to me, as is my illness. Without anxiety and illness, I am a ship without a rudder . . . My sufferings are part of myself and my art. They are indistinguishable from me, and their destruction would destroy my art."

I knew a great musician who once said that he plays not because he wants to but because he has to. On the other hand, there are many creative individuals who have pushed past these struggles without losing their creative ability. Creators like director Aaron Sorkin have described the same fears of losing their creativity with sobriety, but have not experienced the loss they feared.

Environments of strife and oppression can also often be fertile ground for rich creative works. There was a time in the West Indies when the British tried to ban conventional drums and drumsticks. The instruments were taken away, but in response, a subjugated people found a way to prevail. They took old steel containers and found a way to use them as instruments, first without much modification, but later with careful tuning and innovation. By pounding little grooves into the bottom of various sizes, they were able to get different regions of the instruments, called steel pans, to vibrate at different frequencies. If you visit the West Indies today you will likely at some point be treated to the amazing music of a steel band orchestra. The sound that they achieved is truly out of this world. These resilient people were propelled by the great pain and injustice they faced.

Was the creation of blues, jazz, disco, funk, and rap music similarly aided by the inordinate pain and suffering of the African American people in the US? How else could a people who make up just over one-tenth of the population and live under such harsh conditions create such a large percentage of popular music? I believe that there is a connection between people who have nothing to lose and incredible artistic achievement. One might think nothing good comes from anger, pain, and frustration. Although I agree that most decisions made out of anger, frustration, and pain are bad ones, I think perhaps a huge amount of great art is fueled by these emotions. It begs the question, Do art and creativity truly thrive when there is no anger and strife driving them?

For many of us who lack these monumental motivations, who are not faced with such great pain and challenges, creative drive must come from somewhere else. But where? Why do we drive ourselves? What pushes a person to play the guitar until their fingers are raw? Does a painter with all kinds of other fun things to do in her life spend 15 hours a day for days and days perfecting the ability to capture a landscape on canvas?

Intrinsic motivation is frequently associated with creative endeavors. This is the kind of motivation tied to the enjoyment of the task itself, the desire to create, and the feeling of accomplishment that is achieved when a work is complete. Some creative individuals are motivated simply by the act of creation and the resulting sense of fulfilment or wonder. Extrinsic motivations such as a drive for wealth and praise on the other hand may sometimes hinder creative thought, but are still powerful motivators. Creators that seek recognition and riches or immortality in the form of a grand legacy may find these extrinsic factors are what drives them. The desire to leave an impact and change the world is certainly a powerful desire. Every creative individual has their own blend of motivations that puts them somewhere on the gradient between being extrinsically and intrinsically motivated. Even so, someone on the outside may never be able to fully understand all that motivates an individual. The person themself may never know completely what drives them. Some motivations though, like fear and love, are motivations we can all understand.

FEAR
Fear can play a very important role in the creative process, yet it can be a double-edged sword. Sometimes fear is a get motivator because you may be desperately trying to find a solution to a problem that you know will eventually kill you. If you are afraid of dying or being alone or losing something very valuable the fear will help you to focus and give you a reason to do the hard work that creativity can require. The fear can make you keep at it even after you have failed at something again and again. A lot of creative innovations were prompted and motivated by the US competition with Russia for the domination of space. As soon as confident Americans, fresh from World War II, saw that there was a Soviet satellite called *Sputnik* flying over their heads they shifted their creative apparatus into high gear. Certainly the original motivation to create fighter jets that could go twice the speed of sound or fly at heights that approach the top of Earth's atmosphere was the threat of war and nuclear annihilation.

But fear can be crippling. If you are afraid that when you put in the massive amount of effort it can take to create something unique it won't pan out, the fear can discourage you from trying hard. If you have been told and taught all your life to be risk averse and keep your head down and try not to be out of the ordinary, then you may be reluctant or even afraid of doing that out of the ordinary creative thing that you had in mind to do. Fear can be sewn deeply into the environment that you are raised in. If you grow up surrounded by people who have been victimized it can produce an extremely crippling effect. When you're circling the proverbial wagons, it's very difficult to look outside of the box that society has created for you and dream of something better. When many people who live in close proximity to you have been traumatized, very negative messages can affect you like a virus.

It is apparent that people who have had a lot of success can sometimes find it easier to commit themselves to tasks that they perceive will be challenging. Past successes can engender a "winner's attitude" that permeates many other things they do that are not related to the original success. When you know deep down that out of all the people who were trying, you were the one that won that swimming championship, it can give you the feeling that success and progress is the natural arc of your life. It certainly helps you to visualize what it will be like when you have adopted a new idea.

A new idea can require a new way of thinking. A new way of thinking can mean examining the truths that you have held dear for a long time and challenging them. It is very human to listen to evidence that supports what we already believe and discount obvious evidence to the negative. This is what is known as the Confirmation Bias. No one wants to admit to themselves that they got things wrong because it means that they may then have to question all the other beliefs that they have held dear. It may mean that they have to admit to many other mistakes that they might have made. It's a lot of work. So why do it? We do it because it's great, and it sets you free. There's a song called "You Gotta Be" sung by Des'ree. It's a beautiful song whose lyrics remind you of what you need to be in life: Be bold, be hard, be stronger, be calm, be tough, be cool, and be wise. It's all true.

RESPONSIBILITY

Perhaps the purest motivator of creativity is love and faith. When you give your life to serving others and devote yourself to solving problems for others, it can be an amazing motivator. It can send you on a challenge that will never end but will also be endlessly rewarding. You can set all sorts of lofty goals and enjoy an entire lifetime of achieving them. Imagine being the inventor of a device that inexpensively and permanently corrects childhood diabetes. Imagine that you held the patent and you somehow had the ability to set the price. Imagine that you know if you set the price at ten dollars per therapy you would end up making about one million dollars each year. That would be the dream, right?

Now compare that to a different invention. Imagine that you were the person who invents a device made for the battlefield: a small landmine. The design is efficient and dangerous, sporting fins that guide it to the ground after it is dropped. It's brightly colored and looks benign at first glance, a lot like a toy, in fact. So a passing child may spot one, pick it up, and play around with it, none the wiser. This mine might not kill that unsuspecting child. It's not that powerful. Instead, it might destroy their hands, wounding their chest, and maybe blinding them when it goes off. Let's say this is a design that would get you ten times the revenue of the diabetes cure. How does that feel? Is any amount of money worth such an unconscionable design?

As you may have guessed, this is not just a hypothetical design. What I just described is a Russian PFM-1 mine, a "butterfly" mine, which has been scattered en masse across Afghanistan, causing a staggering number of injuries, almost solely to children.

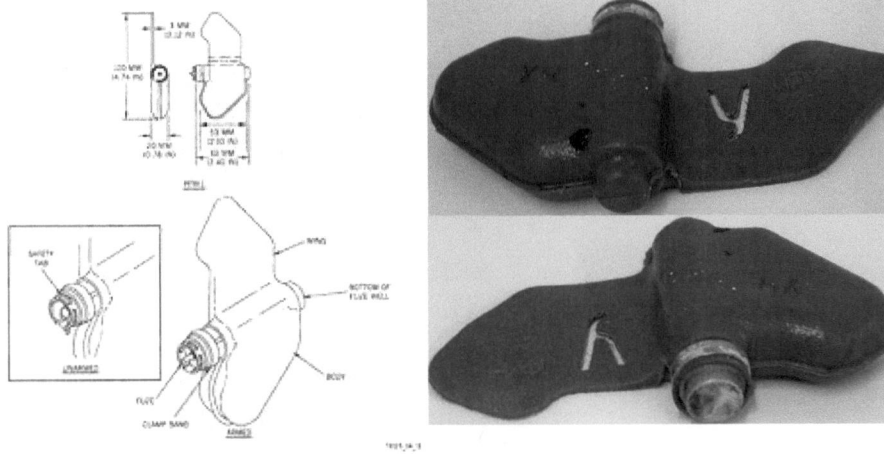

FIGURE 1

It may be hard to imagine creativity in such a bad light. I mean, when you think of that word, what comes to mind? Beautiful art? Maybe your favorite book or video game? But creativity is a tool of the human mind, which, as we can see every day, has the ability to produce both incredible beauty and unfathomable destruction. Humans have a long history of both, but the bad side of creativity, negative creativity, is often neglected in discussions in the subject. Maybe this negative creativity manifests as a student working feverishly to figure out a way to cheat on their final rather than study. The ways we humans find to slide past justified regulations and invent ever more efficient ways to kill each other are all examples of this flipside. Even well-intentioned inventions can turn horrifyingly wrong. The researcher who discovered the foundations for what would become Agent Orange set out to help soybeans grow faster. Instead, his legacy became one of incredible pain and destruction. On the subject, he is reported to have said, "Nothing that you do in science is guaranteed to result in benefits for mankind. Any discovery, I believe, is morally neutral and it can be turned either to constructive ends or destructive ends."

There is a certain responsibility in creativity. It's not only in doing your due diligence to make sure your design doesn't fail and is safe to use, but also in what impact it will have on the world. Remember that your designs, what you devote your creative efforts, blood, sweat, and tears to, do not exist in a vacuum. If they did, there would be no point. Keep in mind what you are devoting yourself to. What impact do you want to have? These thoughts will drive you to create and ultimately influence what you create.

Sadly, our earlier positive example, the diabetes cure, is far from being realized, but beautifully beneficial designs are not unreachable. When you are able to imagine the joy of the prospective users of your product or invention it can keep you going. Some of the best inventions are the result of folks who are in close contact with those in need. A pair of South Africans, aware of the needs of those who have to travel for very long distances to get water before carrying it back, designed an elegant solution in the form of the Hippo Water Roller (https://www.hipporoller.org/). It's a water tank designed in such a way that you fill a tank that functions as the "wheel" and push it like a lawnmower. It allows the ground to bear the weight instead of your shoulders. I've never used the invention myself—I live in the Silicon Valley—but I imagine if I had to get water from a few miles away I would really be thankful for one of these. This is a product that provides a huge, positive value to its users' lives. Imagine how it would feel to implement such a design. Pretty good, right?

FIGURE 2

HARNESSING CREATIVITY

We know that creativity is essential to inventing, but nailing down your own creative potential is a hard thing to do. How can you enhance and harness your creativity to come up with great ideas? It often seems like creativity is something random and uncontrollable. Creative insights can come seemingly out of the blue and even those who have studied creativity for decades can't account for everything that goes into creativity and the bud of a new idea. Whole books, journals, and fields of research are devoted to answering that very question. However, there are some tried-and-true methods people use to get those creative juices flowing. Here are just a few recommendations to get you on your way.

ACTIVE STRATEGIES

A certain state of mind can encourage creativity. There are a variety of traits that have been shown to be conducive to creativity. Some of these may seem like no-brainers. Creativity is enhanced when traits like openness to experience, flexibility, autonomy, and intrinsic motivation are present and cultivated. An environment that is tolerant, diverse, and multicultural, embracing a variety of perspectives, is conducive as well. Creativity is crippled by the opposing attributes such as rigidity and closed-mindedness. Mood has an interesting effect as well. Positive moods tend to encourage wider associations and risk-taking that is valuable to creativity whereas negative moods can lead to higher levels of persistence but more narrow thinking.

In general, strategies used to encourage creativity fall into two camps: active and passive. Active strategies are the ones that you "make happen." Frequently, these involve a conscious shift that facilitates "out of the box" thinking. Brainstorming is perhaps the most well-known example of a creative tactic, but many others exist.

For example, it is often helpful to change your perspective. Looking at a problem from another point of view or turning it on its head is often conducive to a breakthrough. This is one way that outsiders to a field can do so well. They aren't weighed down by the assumptions and similar perspectives of those who have worked in that field for decades, and as such can often offer fresh insight. Their ideas may frequently be novel but useless, but they also have the potential to make breakthroughs no one else would consider because of their unique perspective. AutoTune, for example was made, not by a singer, but by a petroleum engineer. People often recommend travel as a way to change perspective, but sometimes that is not possible to the extent we would like.

Reframing the problem itself, changing how a question is posed, or altering the focus of inquiry can be helpful as well. Zooming out or in in scope can help you make connections to a bigger picture or isolate a small part of a complex issue. Think of a jigsaw puzzle. Sometimes you have to look at the whole picture to make any progress, but other times it is better to examine the attributes of just one piece to see where it fits. Changing the representation of a problem can be helpful as well. This can be something as simple as drawing out an idea that's just in your head or using a map to express written or verbal directions. Each of these methods can unstick a mental roadblock or open up new avenues that would be hidden if you're stuck in one way of seeing something.

Another useful tactic is using analogies and analogous relationships. Many creative ideas take an aspect of one item or concept and combine it in new ways with something else. Steve Jobs used the metaphor of an actual desktop, with stacks of papers and the like to model the interface of Apple computers for ease of use. Consider also the famous example of George De Mestral's invention of the hook-and-loop fasteners he named Velcro after he noted the way burrs stuck to his dog's fur. It's not uncommon to find inspiration for a solution or design from similar or even seemingly unrelated examples that already exist. Nature is full of opportunities for analogous applications and ideas, and is an amazing source of creative potential. Even spending time in nature has been identified by many creative individuals as a great way to encourage creative thought, especially as the mind wanders.

Some active tactics can be extremely specific to a field or an individual. In design processes, there is a technique called attribute listing. It is a systematic method by which you identify all the essential components in a product and go through each one, identifying the myriad ways it can be resized, recolored, made of another material, or otherwise changed. Great creators tend to have their own distinct creative processes. One paper distilled Edison's inventive process down to six steps: "Define the need, set a clear goal and stick to it, analyze the process and stages involved, assess objectively the progress, keep each team member on task, and record the work for possible examination at a later time."

Sometimes putting yourself in a certain environment is the best approach. Another way of finding the best idea is to have two or more competing teams. Naturally this is very expensive unless you incentivize the teams to move forward. The drawback of this approach is the cost of having so many teams involved. If you create a contest among universities or other types of professional organizations this could serve as a viable way of finding your idea. Of course, for some people the pressure of competition drains their creativity. It all depends on the individual.

PASSIVE STRATEGIES

Sometimes, however, it seems creativity acts like a lightning strike, a flash of brilliance coming out of nowhere. This is often a result of the uncontrollable side of creativity, in which an idea incubates in the subconscious only to burst forth apparently fully formed like Athena from Zeus's head. This is where passive tactics for encouraging creativity can come into play. Have you ever gotten a great idea after going for a walk or exercising? What about in the shower or just before you fall asleep? These moments in which your mind wanders or goes momentarily blank are fertile fields for creative ideas. Taking a break from a task may seem counterproductive but can help provide the insight you may need. It may not even be necessary to take a complete break. This ties well into the way that switching tasks can help you avoid getting stuck in a rut. It may seem as though all of your focus has shifted away from the problem at hand, but often there is something going on under the surface that will bring you the solution you need.

Even still, sometimes it all comes down to random chance. Math and mathematical techniques are something that exist whether humans find them or not. At any time some incredible visionary, without the help of a research group, super computer, or multimillion dollar lab, can uncover this method of doing things that can change the world. I believe that a good design has the same property or perhaps is just a manifestation of the same exact underlying truth. A modern bicycle is an amazing invention with a main advantage that comes from the main component, the wheel. The wheel exists in nature as anyone who's ever thrown a log underneath something to move it along knows. It's a pretty sure bet that the first human to use a wheel or wheel-like device was not alone. Had he or she never been born, surely someone else would have been the first and probably in a similar time frame. I imagine that the first use of a wheel was accidental. Perhaps someone was sitting on a log on a hill and it started to roll or something like that. Sooner or later someone thought to try it again and do it on purpose.

I think a lot of inventions and discoveries are like that. An accident allows a certain insight. The incident is recorded and repeated. The method is refined and put to use. If you are trying to invent a new handheld device, perhaps the next iPod or Kindle, I think it's instructive to know that the actual design exists already, we as inventors and designers just have to uncover it. Sometimes it's simply the convergence of a perfect storm of attributes that brings about some innovation.

It may seem counterintuitive, but simply switching to a different problem is also an effective way to encourage insights. Creative individuals often have multiple projects going at the same time that they can switch between. It is important to note that changing your focus is not the same as multitasking, which, instead of allowing one problem a little mental "breathing room," splits your focus and hinders creativity with distraction.

KNOW THYSELF

All of these tactics are well and good, but it is also important to be self-aware. Take note of your assumptions. Everyone has them, and they often serve a beneficial purpose of freeing up cognitive load, but they tend to work against creativity. Notice them and question them.

Find the tactics that work best for you. You may notice that you do some of these things already when trying to come up with an idea. You may find a rigid process like diagraming works best for you. Likewise, you may learn that a more free-flowing strategy like a nature walk works even better. Don't be afraid to try something new. Creativity thrives on change. Combine techniques and be flexible. Avoid getting stuck in a rut. Just because a tactic works for you doesn't mean it's the best one for every problem. And remember, this is not the end-all be-all. We learn new things about creativity all the time. Just keep an open mind, keep learning, and keep creating.

PERSISTENCE

A certain part of creativity can be just hammering it through. When you've had that initial creative idea it's often not enough. Sometimes the idea needs to be refined. Sometimes the initial idea is really quite lousy. In these cases nothing will come of the idea, but thank goodness there are people like my mom. Evelyn Samuel was a nurse who became a teacher so that our family could all have summers off together. She was and is the kind of teacher and person who cares very deeply about others, but who never lets you off with an excuse. Discipline and hard work are mandatory. As I grew up, and to this day, Mom was and is a powerhouse. She cleaned the house to a fine finish, did the cooking and all the housework, and was always sewing or working on her two masters degrees. I knew she was an awesome force to be reckoned with, but on some level I thought it was normal to be so strong. My mom would do the right thing and pursue it hard. If she's in the room, everyone can hear her voice. When we built additions onto our summer shack, she would pound in nails just like Dad, Mike, and me.

My mom used to take me to Cub Scout meetings, and one Halloween we had a costume contest. Mom and I designed and sewed a costume together from scratch. Mom taught me how to use the Singer sewing machine that was one of her wedding presents—I own it to this day. The costume was a large mouse with a long tail made out of green silky material. I was so proud to wear it and so proud of Mom and myself when we won first place in the competition. My mom makes excellent quilts to this day. There is an old saying about quilting, "To start a quilt is admirable, to finish, divine." I think that sums up my mother's personality and the effect that she's had on me very well. She does what she does and finishes it. She puts her head down and accomplishes things. This is a very important element of creativity. Without the power to fuel a creative idea to its end nothing moves forward.

Persistence is key. It has often been identified as a hallmark of creative individuals. How many inventors can you recall that have tales of seemingly endless failures before their ultimate success? Studies have shown that it can take quite a few mundane ideas before reaching a truly novel one. The more ideas generated, the more creative they tend to be as time goes on. As Thomas Edison is famous for saying, "Genius is one percent inspiration and ninety-nine percent perspiration."

RECORD, RECORD, RECORD

Don't let good ideas go down the drain. One can only imagine how many truly great ideas have been lost because they were simply not written down. Always be ready when inspiration strikes, whether you're in the shower, about to fall asleep, or on your commute to work. It is definitely possible to have a great idea and then lose it when you get interrupted or move onto something else. Keep a notepad or a small tape recorder with you at all times, or make effective use of your phone, which can usually function as both.

BARRIERS TO ADOPTION

In the first 18 years of my life growing up in close proximity to people that were doing very poorly economically and socially, I noticed that some of the biggest, if not the biggest, barriers to progress and growth were attitude, mistaken ideas, and disadvantageous habits. I've come to believe that human behavior has a stationary momentum of sorts. Most of us wander through life with certain assumptions and ways of doing things. It doesn't matter if these assumptions and ways of doing things bring about obvious pain and suffering, they're a pattern that is followed regardless. As an African American youth in the '60s and '70s, growing up on the streets of Brownsville, an impoverished neighborhood in Brooklyn, NY, I've seen people act impulsively and, in an instant, destroy their entire future just because they were taught the idea that it's okay to do just what you feel in life as soon as you feel an impulse. I've seen fights break out just because one person looked at another the wrong way. We've all seen news stories of people shooting each other over something as insignificant as a parking space. These are extreme examples of people locked into a way of thinking, but they illustrate the basic principle.

The way this principle relates to inventions is the fact that when you come up with a great new idea, one that may save time and effort, it won't always be embraced just because it's new. It may be rejected or not even tried just because it doesn't comport with the tried and true method. I've witnessed a strong ethic in certain regions of the US where folks have a heightened and admirable love and respect for their ancestry. However, the ethic is sometimes taken too far in that there is a feeling that the way that you've always done something is the way you should continue to do it out of respect for your elders and your ancestry. When you come up with a new idea or a new way of doing things you run the risk of going against some learned value. As I was growing up there was a saying that you shouldn't wear white after Labor Day. To this day I have no idea where it came from or who said it first, but people followed it. I've always felt that it was utter nonsense. I've always been atypical. The neighborhood of my young life was filled with these assumptions and baseless ideas and for many reasons I was spared their overwhelming negative effects. I was able to leave the neighborhood every summer and see and be exposed to many people who felt very differently in many ways. I learned time and time again that no one people or culture has all the answers.

There are certain industries and products that are traditional to the point where new products will be rejected without thought. One of those places is the bathroom. For a very long time we've had the same toilets that still get clogged every now and then with no button or function that automatically unclogs them. There are very few toilets with a two-mode flush to save water. Toilets still don't have technology that would stop odor from reaching our noses. It would be incredibly easy to fit each toilet with a device that would create a minute negative pressure in the toilet cavity when a person sits and route the air to a HEPA filter. In most cases bathroom stall doors open inward so when you want to exit the stall you have to do this incredibly awkward dance around the bowl to get out of the way of the door as it swings inward.

Why doesn't a sink flush like a toilet? Most people who brush their teeth over the sink or shave get stuff all over the inside. Then when you want to get rid of all that stuff many of us cup our hands to catch a bunch of water coming out of a little fixed spigot and pour it all over the places in the sink where the water doesn't reach. It's time consuming and dumb. There are still many showers with a curtain that begins to come toward you as soon as the hot water comes on. There's a nice fix for that, it's a shower curtain rod that bows out away from the shower enclosure but as of the writing of this book there are still plenty of showers with the old-fashioned curtains and rods —yuck!

The point is, the world is ripe with areas for improvement that people simply don't act on. There is risk in any innovation and sometimes the tendency to adhere to the status quo will prevent an idea from getting far, but a good idea that improves the lives of its users will speak for itself in that respect.

THE GUTS TO CREATE

So what does all of this mean for you? Ultimately, the takeaway is this: Creativity is an essential part of the entire process of bringing an idea to fruition and there are many ways you can try to enhance it. It is important to note though that creativity alone is not invention. Invention requires implementation, thorough design, and all the steps it takes to make your idea a reality. Invention, which is the true focus of this book, consists of all the steps that follow after the idea. An idea is nothing without investment. Sooner or later the idea has to be reduced to practice.

For now, congrats! You are part of a very special group of folks who are creative enough to have your own idea for an invention and confident enough to actually do something about it. You have already jumped the biggest hurdle—many people have great ideas that they never move on, only to see "their" idea on the market later from someone else. Now that you've made it this far, there are a few things that you need to know to reap the benefits of your idea as soon as possible. This book aims to cover them all.

PRODUCT DESIGN

A PARADIGM FOR INDUSTRIAL DESIGN

The basis of a lot of good design is the principle that *form follows function*. When the form of a product is closely related to its function, the design will be optimized. It will be honest and won't have a lot of extra shapes, superfluous curves, or wasteful geometry. Many think that a design appears cheap and junky when geometry looks like it's thrown together without reason. This is true, but it's not the only element that guides good design. There is something more fundamental to consider: the principle that "form follows emotion." This saying was one of Steve Jobs' guiding principles in the design of Apple computers. A great design begins with the emotion behind the function, which will inform the final appearance of your design.

If we start with the emotional need that the function fulfils we are on the road to a great design. Let's say we need to design a meter to ensure that miners don't get killed by exposure to poisonous gas. Such a meter has to be worn on the outside of their clothing, and the technology it contains converts the presence of trace elements of poisonous gas into electrical signals. The signal causes the meter to initiate a number of alerting behaviors. It beeps, vibrates, and lights up. But what is the emotion behind this product? Fundamentally, it's the desire to stay alive. This informs the nature of the device. It is a protective device, and it is protective in a communicative way. It suggests, rather vigorously, that the user leave the area occupied by poisonous gas. It's not a shield, but rather the canary. It's a device the user must be able to trust to inform them of danger so that they can react and, ultimately, survive.

Understanding the function on an emotional basis allows higher quality decisions to be made about the form. The functional elements of technology can be packaged into something that is sleek and lightweight, yet sturdy enough to survive any rough treatment it might encounter. The choice of a surgical steel exterior will communicate to the user that the meter is expensive, sensitive, and high tech.

On the outside there will be plenty of warning beacons that light up and a place for a buzzer. There will be bumpers on the product so that if it gets hit while a person is working, the meter won't be damaged. The covers over the lights will be sturdy, shock resistant polycarbonate. There will be fins in front of the vent holes so that if the user places the unit in a pocket, the fins will create a little breathing space by holding the cloth away from the breathing hole. The clip on the back of the device will be as slim as possible so when it is hooked onto the top of the pocket it doesn't hang down at an angle. All these decisions fit together to serve the basic function of the device, but more importantly they need to evoke emotions tied to safety, protection, and wellbeing. These emotions must be hardwired into the physical structure and workings of the device.

In the case of Apple, sticking to this principle has brought incredible success. When creating the design for the 1984 Macintosh computer, Steve Jobs insisted on an adherence to the emotion behind the form. He wanted the computer to be "friendly," a description that baffled his employees and competitors alike. His vision soon began to take shape though, and that friendliness manifested in a design with light colors, curved edges, and a form that resembled a face. The approachable nature that drove the design was evident in the language used to advertise it, with one ad calling it "a personal computer so personable, it can practically shake hands." The emotion behind that from, which directly addressed the intimidation many felt at the time when faced with the not yet mainstream personal computer, got right to the heart of what the target users wanted. Modern design needs to go beyond the old adage of form follows function and move instead toward a paradigm that seamlessly integrates emotion, function, and form.

THE STATUS QUO

According to a famous myth, Alexander the Great was faced with the difficult task of untying the Gordian knot back in 333 BCE in the Kingdom of Gordian. The Gordian knot was an extremely complicated knot that was made of hundreds of tightly interwoven thongs of cornel bark. You couldn't even see where the ends were. Many before him had tried the challenge, and all had failed. When it was Alexander's turn, he solved the problem by taking out his sword and hacking the knot to bits. In essence, he solved the problem using a simple, streamlined approach. Instead of following the well-worn path of those who came before, he found a new way that circumvented the challenge of the problem entirely.

When faced with a design challenge, it is often natural to consider the problem in terms of the way others have considered it before. We all have a tendency to lock ourselves into what is familiar. For example, if you were asked to design a new toaster, it would be perfectly natural to go out and purchase a bunch of toasters currently on the market, attempt to figure out what makes the good ones so good, and finally adopt those attributes for the new toaster. In many cases when one spends so much time studying the designs of others it influences us to make our designs very much like theirs, with only slight variations.

A great example of this is nighttime TV. When you consider how much money it takes to create an hour-long television show and how many of the shows are almost identical and follow similar, predictable formulas, it is easy to see that the creators were overly influenced by what has been done before. Such copies that adhere to a status quo often draw from a mentality of safety and minimum risk. After all, if something has succeeded for years, why take the risk to change it?

This sort of mentality is anti-creative, but unfortunately common. The automotive industry probably provides the greatest example. With so many clones on the market it's astounding how many just follow the pack. One would think automotive engineers would be some of the most mechanically creative people on the planet, but year after year we get only small changes until a company like Tesla comes out with a revolutionary electric car.

In the spirit of Alexander the Great, sometimes it's really great to totally throw away the status quo. For example, if you look at the power cord on an Apple computer you will notice a simple innovation that solves a big problem. The power cord is held onto the computer by a series of magnets. It's totally secure, yet if someone accidentally kicks the cord as they walk by, the cord will let go instead of throwing the computer to the ground. It's a stroke of genius that totally goes against the status quo. It speaks volumes of Apple's long-term commitment to hacking away at our preconceived notions and taking us further and further in our thinking and behavior.

THE BENEFITS OF A SIMPLE DESIGN

Another thing that made Alexander the Great's solution so effective was its incredible simplicity. Experienced designers know that the KISS (Keep It Simple, Stupid) method is the one that's time tested and usually provides great results. A simple design is usually more elegant and has fewer moving parts. When a part is eliminated from an assembly, you also eliminate the need to model and draw it. You also bypass levels of manufacturing, inspection, price negotiation, and myriad other things that each take time and effort.

When you try to pack a design with every feature, it doesn't always result in a better product. Think of a young child who receives an amazing motorized toy. These days toys do things that were inconceivable a mere 30 years ago, but in many cases, the toys that get played with again and again, often for years, aren't these motorized spectacles, but the simple ones. Simple blocks, dolls, or even good, sturdy cardboard boxes don't have a lot of features, but they leave open the opportunity for endless inventive play. Similarly, many of the early wireless network devices came with a long list of features and capabilities. Consequently, many were a nightmare to install. You were guaranteed hours and hours of reading through manuals, calling hotlines, and doing frustrating experiments before you finally got your system up and running. Later, wireless network boxes switched to the plug-and-play style. They got simpler. Many of them did a little less, but they were far easier to use. Clearly the benefit of simple design is aesthetic, cost saving, and promotes ease of use.

Still, a simple design is not necessarily one that excludes technological complexity. In fact, the better use you make of complex components, the simpler the product can be in terms of its ease of use. In this case, it is essential to think through the simplicity of the user experience. For example, when you drive down the highway in a hybrid car, a plethora of things controlled by a computer are all happening at once.

There's regenerative braking, there's the transfer of power back and forth from a gas engine to an electric motor, and many other electro-mechanical operations. Yet the result is you get in the car, step on the "gas pedal," and get where you're going with better gas mileage than you would have without. The simple interface hides the complexity under the hood for a simpler and more enjoyable user experience.

THE RUDIMENTARY DRIVING FORCE BEHIND ALL PRODUCTS

When designing a product, an inventor such as yourself must keep in mind what drives the buyer to purchase that product. Often at the heart of these choices are basic, visceral urges like hunger, sex, power, fear, and safety. Know what motivates your target consumer and strive to meet that motivation. Perhaps they are looking for a motorcycle that will make them feel powerful and the center of attention. A quiet, compact electric motorcycle may seem like a great design improvement, but it would leave that customer completely unsatisfied. If they are looking for a car that will make them feel incredibly safe, all the cool designs, bells, and whistles in the world won't matter when you don't even remember to add airbags. It can sometimes be tempting to design for ourselves, but the perfect product to meet our desires may not find an audience among consumers. As we strive to create products that sell we must be cognizant of these rudimentary driving forces. Of course, it gets way more complex in its actual application, but understanding the rudiments is key.

THE ESSENCE OF BEAUTY IN INDUSTRIAL DESIGN

One of the most important roles of an industrial designer working in product design is to define the product's form so it will attract and delight the end user. For the most part, the first impression that a buyer has of your product is visual. They see it on the shelf, or in a photo online, or in the hands of someone else. From that moment on, your product has to work. It has to work visually to convey a powerful message about the product. The message can be, "You need me," or "I'll take care of you," or "With me you'll have more free time," or "You'll get more of what you want." Even huge pieces of industrial machinery benefit by saying something that is visually attractive. They do it with color, shapes, textures, graphics, choreography, and anything else that can be made appealing to the human senses. The form can capture the eye of a prospective customer and beckon them over to the shelf, or entice them to click on the link and input their credit card information.

A product's form can give the user a sense of pride when it's seen by friends, family members, or even total strangers. Form gives the user something that somehow appeals to an underlying sense of beauty and value. Although the function of a product is arguably more important than the form, and the function is somewhat inseparable from the form, it is instructive to consider the form as its own separate entity.

Great industrial design is beauty and art. Unfortunately for the industrial designer trying to give a product a form that will help it sell this can be difficult. Art and beauty vary greatly among cultures, age groups, and individuals. When defining a product that is supposed to sell worldwide, it is hard to know what will be appealing to such a varied user community. That said, industrial designers who are trying to suit the tastes of a huge and varied audience benefit by considering what is universal.

THE BEAUTY OF NATURE

When we look at the natural world we are viewing the art and design of the best designer in existence. Arguably there is nothing that can be created by humans that will ever come close to the beauty of the natural world, but industrial design can be greatly enhanced by borrowing shapes, colors and textures from the natural world. The beauty of a rose, the beauty of a young foal, and the beauty of a majestic bird of prey can aid in the creation of product form.

When certain sports car designers want to define a fabulous look for their automobiles, they may attempt to emulate musculature, teeth, and other physical human attributes in their designs. The designers even attempt to connect their products to the natural world with the names they choose, such as Mustang, Thunderbird, Cobra, Viper, and even Rabbit.

FIGURE 3

Some of the most beautiful and appealing designs appear as if they have grown out of the landscape. In architecture this can be achieved by deviating from the perfectly straight and vertical surfaces that commonly appear in man-made items and utilizing instead the soft sloping curves that appear in the natural world such as in waterfalls, trees, and dunes. It can also be expressed through asymmetry.

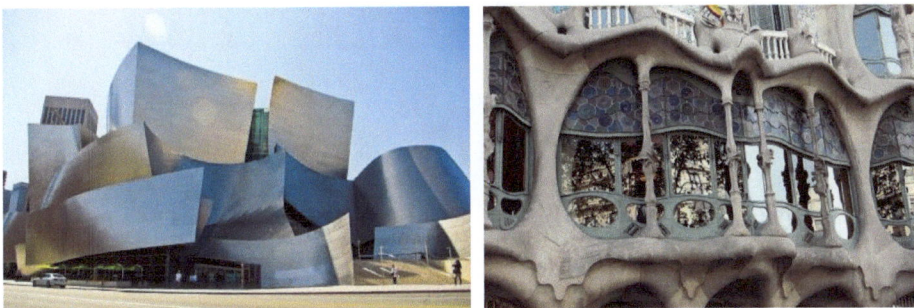

FIGURE 4

The architectural design on the left is by Gaudi and is located in Barcelona, Spain. The architecture on the right is the Disney opera house in Los Angeles designed by Frank Gehry.

The same principle is applied to product design. This ring leverages similar principles of asymmetry.

FIGURE 5

RING UTILIZING ASYMMETRY

When products are created that make use of natural lead-ins and thickened bases, they look stable and rooted to the ground and borrow visual cues from natural shapes like tree trunks and flower stems. The user can feel comforted in the same way a hike through a forest may bring a feeling of calm and sense of beauty.

FIGURE 6

A SET OF PILLARS, A TREE TRUNK, AND BELL BOTTOM PANTS

The raw material of a design can harken to nature, too. The most overt way to do this is to use natural materials in their natural state such as wood and rattan for furniture and other products.

FIGURE 7

RATTAN CHAIR, WOOD SCULPTURE, AND PHONE CASE

THE HUMAN FORM

The human form can be an incredibly compelling tool of inspiration for product designs. There is something instinctive about human reactions to these forms that is pleasing and draws our eyes. The human body is seen by many as a pinnacle of design and has been revered in art for centuries. In product design, elements of the human form can be incredibly beautiful and attractive to users.

FIGURE 8

SCULPTURES OF DAVID AND VENUS

Humans are hardwired to recognize and connect with faces. It makes sense: Faces are the primary way we identify and communicate with others. Faces bring up instinctive emotional reactions based on expressions and features we recognize automatically. Humans see faces everywhere, even in something as simple as the arrangement of three dots. Who hasn't looked at an electrical outlet and seen a surprised expression looking back? Designers can take advantage of these instinctive reactions by incorporating elements of facial features into their products and tailoring them to fit the personality they want their design to have. This is a technique often seen in the front headlights and grills of cars.

FIGURE 9

THIS VW BUG GIVES THE IMPRESSION OF BIG EYES AND A SMILE, CONTRASTED WITH A MORE AGGRESSIVE-LOOKING FERRARI.

FIGURE 10

MORE SUBTLE IMPRESSIONS OF FACES CAN GIVE A FRIENDLY APPEARANCE, SUCH AS WITH THE 1984 MACINTOSH.

The rest of the human body is also a great source of design inspiration. Such inspirations can manifest as impressions of toned musculature, lithe curves, and other attractive elements of natural forms. The female form is particularly compelling. Humans are naturally attracted to soft, natural curves like those so celebrated in the female physique. The female form can also bring up emotional associations, calling to mind the nurturing compassion of a mother or alternatively feelings of sexual attraction and arousal. The appearance of women is often used in advertising to attract people to products, but more importantly, it also influences the form of the products themselves. Some industrial designers will even reference pictures of beautiful women as they begin sketching lines of a radio or a computer. Sometimes this approach may lead to extremely blatant visual cues, but in many cases a more subtly influenced design can produce a fabulous and widely appealing one.

FIGURE 11

OVERT IMPRESSIONS OF THE FEMALE FORM MADE VISIBLE IN THESE CHAIRS.

BEAUTY FROM INCREDIBILITY

In some cases, the beauty of a form is derived from an incredible and sometimes supernatural or futuristic nature. When a consumer sees something that looks impossible, it is attractive. For example, lighting fixtures or displays that seem to glow without an apparent bulb, or one-legged tables that seem to stand on their own appeal to a sense of magic and wonder in us. How do they do that? We simply are attracted to things that are of a high-tech nature or are advanced or astonishing in some way.

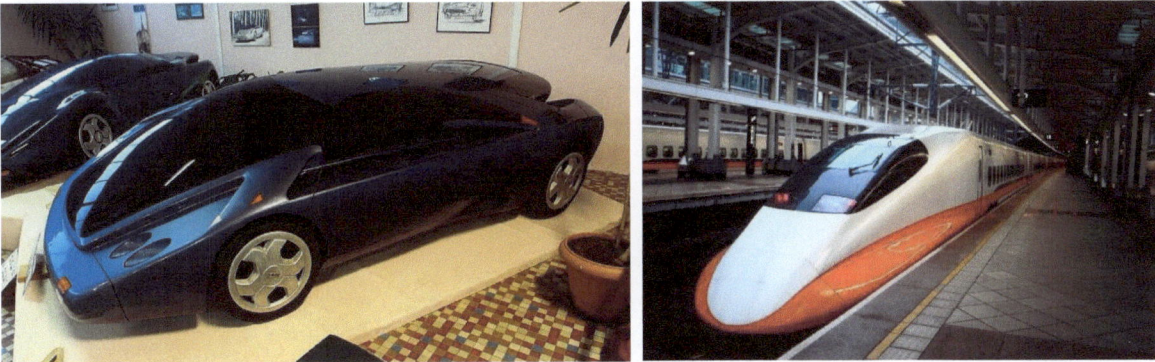

FIGURE 12

DESIGNS OF FUTURISTIC MOTORIZATION.

However, anything that derives its beauty from wonder runs the risk of becoming commonplace when the novelty wears off.

Some products are amazing because of how simply miraculous they are. When you see a newborn for the first time and you look at the little feet with the perfectly formed little toes, it's incredible to think how they have all the same blood vessels, bones, and nerves we have, but in an incredibly small and perfect package. The look of a pregnant woman can be amazingly beautiful not just because of the curves and shapes, but largely because of the amazing miracle that's going on inside her.

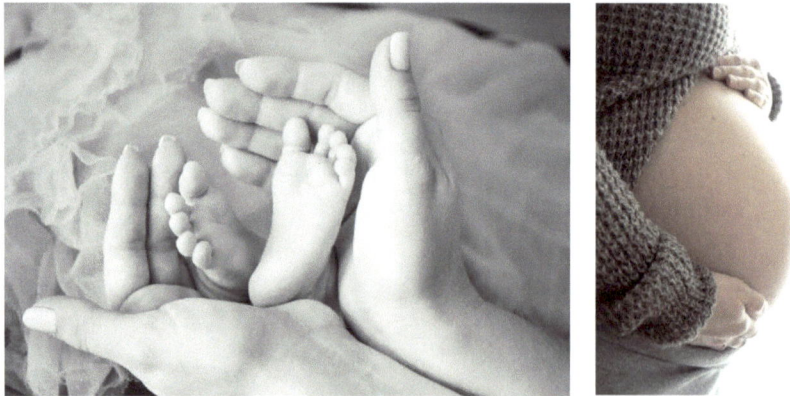

FIGURE 13

SOME OF THE GREATEST SHAPES KNOWN TO HUMANKIND

These principles also apply to good product design. When certain Apple laptops are manufactured, they are machined out of an aluminum block. This allows them to be super thin and achieve a look that's unlike any other laptop on the market. The idea of machining a computer chassis out of a solid aluminum block is insane, unless you have the inventiveness and initial capital to set up a dedicated and automated factory to do it. When you're Apple, anything is possible, and the designs are amazing. Another example of this is the iPod. When everyone else in the industry was making the same old injection molded housing, Apple designers decided to use an aluminum extrusion, giving the iPod its unique, polished look. The incredible, compact display makes the user think, "How did they make a display screen that small? It's a miracle."

FIGURE 14

THE ASTONISHING IPOD NANO.

Detail makes a product look amazing. If you compare the paint job on brand new BMW—any model—to that of any car that is half the price, you will see a marked difference. The paint on a BMW is smooth as glass and stays that way even when you leave it out in the hot sun for years. How do they do it? What do they use? It's astonishing.

The color of flowers is arguably more vibrant than anything else in nature. As you walk through a beautiful forest you can be overwhelmed and dazzled by all of it. When you walk up to an amazingly colorful plant that is growing naturally, it's equally astonishing. A rainbow has the same effect, and industrial design is unsurprisingly heavily influenced by these colors of nature.

FIGURE 15

BEAUTIFUL FLOWERS AND A RAINBOW

BEAUTY FROM EXTRAORDINARY HUMAN EFFORT

To some, evidence of extraordinary human effort is itself beautiful. It's very similar to the beauty of incredibility, but there's more emphasis on raw human energy. Consider, for example, the Great Wall of China. It's not really a mystery how it was built, but it's considered beautiful because of its length and the amazing amount of human effort that went into building it. Castles are considered beautiful both because of the massive stone work and because of the sheer difficulty in producing anything like them, even in modern times. The mansions of Newport, Rhode Island, are considered beautiful for similar reasons.

FIGURE 16

FIGURE 14. THE GREAT WALL AND A CASTLE

When every square millimeter of a huge room is covered with some ornate art that some craftsperson worked half their life on, many people find it to be extremely attractive. Some of the beauty comes from the fact that the marble is imported from Italy, the rugs are imported from Persia, or the glass of the chandelier is made from 5200 individually hand-blown glass beads. The curtains may be made from a silk that only grows in one small province of China by special silkworms that only produce eight grams of the stuff each year—you get the idea. Industrial designers can use this technique when considering what materials, processes, and appearances can make their product feel particularly special. This may involve something that can't be mass manufactured or attained by a machine, but instead simply has to be done by hand.

FIGURE 17

ENCRUSTED AND FESTOONED PRODUCTS

FIGURE 18

CARVINGS IN A TABLE AND WOOD CEILING

When you see intricate carvings it's easy to imagine that we are seeing the entire life's work of one person. It may make us feel special that we can walk by it and enjoy it knowing that a person somewhere dedicated years of their life just to produce this one piece of art. It may make us want to learn the means by which to make something like it, or at the very least it may make us more inclined to want the product.

BEAUTY FROM EXCLUSIVITY

When you see items that, if you possessed them, would instantly place you in a special category, they are attractive. Arguably a lot of the "beauty" of these items comes from their exclusivity. The Hope Diamond has only been owned by a handful of people. If suddenly there was a discovery that made it possible to take some sugar and some skim milk and microwave it for 30 seconds to get an object that was identical in every way, shape, and form to the Hope Diamond, would the copies be thought of as beautiful, or would you see tons of them in the landfill?

FIGURE 19

THE HOPE DIAMOND

Have you ever noticed, the bigger the rock, the more people say it's beautiful? It's common practice to cover things in gold leaf. Paint won't do. As soon as you put gold on anything you increase its monetary value, and with that you decrease the number of people who can afford it. Designers can easily take advantage of the exclusivity principle to make their products more appealing. They may incorporate expensive and hard to obtain materials or use techniques that are exclusive or proprietary. For this principle to work well, there must be a reason why the product is not available to the masses. The product might be produced in a limited quantity even though it may not have to be. The product might bear the name of a celebrity.

BEAUTY FROM THE RELIEF OF PHYSICAL NEEDS

Some of the most powerful human desires such as hunger, sex, and safety attract us to that which promises to deliver these things to us. In the art world we find myriad still-life paintings involving, not surprisingly, food and nudes. In this day and age, with a fast food restaurant on every corner, it's difficult for many to imagine what it's like to not have access to good food. Although some would argue that these days good food is still scarce, the perception is you can get good food anytime you want. The painting shown below is typical of many works done in a time and place when fruit was not readily available all year round.

FIGURE 20

STILL LIFE WITH APPLES AND GRAPES BY CLAUDE MONET 1880

Industrial designers can use this principle greatly to their advantage. The choice of color and shape can be borrowed from food. Of course beautiful women are the subject of art and design for many reasons. Beautiful nudes can be seen as the promise of relief from sexual desire.

FIGURE 21

THE THREE GRACES BY PETER PAUL RUBENS, 1639 AND BATHER ON A ROCK BY PIERRE-AUGUSTE RENOIR, 1892

The desire to be powerful and safe can be represented and projected by images and shapes associated with strength. For that reason, products that project a feeling of strength can be very appealing. Products that appear built to last and are associated with strength and toughness can do very well.

FIGURE 22

WAR POSTERS

In a recent automobile TV commercial, a football player is getting out of his football uniform so the audience can view his perfectly athletic physique. Then he is dressed in a fine business suit. In the very next scene a car is shown as if to imply that underneath the elegant design of the car there is a powerhouse of an engine—a real performer. The typical color of guns and knives are dark neutrals like black and gray and dark brown. Products associated with strength tend to be traditionally masculine and in some cases, phallic.

FIGURE 23

RAMBO AND THE CADILLAC CTS

The feeling of safety can also be portrayed with images that are serene. What can be more serene than being in a garden of flowers? Industrial designers commonly borrow shapes, colors, and textures from natural visions of serenity.

FIGURE 24

BED OF CHRYSANTHEMUMS BY CLAUDE MONET

SIZE MATTERS

The size of a product, whether it is small or large in comparison to other things can work to make a product appealing. In each case the size of a product has to indicate that it's special and therefore has increased value and appeal. When products are miniaturized consumers get the feeling that special care was taken to make the parts smaller than normal. The idea is that the product is now more useful, required more engineering, or is more exclusive. In essence, it's more miraculous. You can just hear consumers question, "How did they get all those components into that little package?"

In psychology there is a theory called Steven's Power Law.

$$\psi(I) = kI^a,$$

FIGURE 25

STEVEN'S POWER LAW

It is said to be the mathematical relationship between the magnitude of a physical stimulus and its effect on your perception and reaction. Some psychologists use it to prove that when an object is bigger there is a tendency to place more value on it. Of course here in the US there are a lot of restaurants that attract us with a super-sized meal. We are much more attracted to huge automobiles, boats, homes, etc. compared to the rest of the industrialized world. There are places in the US where men are expected to drive trucks—large trucks—whether or not they have a real need for one.

Some studies have shown when vacuum cleaners are loud consumers assume that they are more powerful. A good industrial designer could easily look at the experience of using a vacuum cleaner, conclude that it's too loud, take steps to quiet it down, and then lose sales. There are even products that are purposely weighted down to add a sense of value. The conclusion is this: Sometimes with certain products, in order to make things more appealing, it's good to make a product bigger than its competitors. Other times a product has more appeal when it's miniaturized, but in general, a product should be as large as the actual function would mandate. For example, the size of a hair dryer should be dictated by the handle to hold it, the diameter of the blower, and the space to house the motor, PCB, fan, and switch. The size of a product is a judgment call.

BEAUTY FROM MATH AND BALANCE

When things are mathematically balanced and clever, they can be very appealing. Many ancient roman arches, modern buildings, handheld devices, and other products make use of mathematical relationships. The golden rectangle is a form where the height and width of a rectangle have a 1.61803399 relationship. It has been theorized that various mathematical shapes compel your eye to move through the shape in a particular pattern that is soothing.

FIGURE 26

THE GOLDEN RECTANGLE SEEN IN THE PARTHENON AND UN BUILDING

The golden rectangle or ratio was used to develop the proportions of the United Nations building as well as the Parthenon. Other mathematical relationships are appealing such as the math behind the domes built by Buckminster Fuller. The calculations that were made in order for the Brooklyn Bridge to stand are inherent in its shape and feel, and it is considered one of the most beautiful manmade structures ever built.

FIGURE 27

FULLER'S GEODESIC DOME, AND THE BROOKLYN BRIDGE

Art Deco is frequently characterized by a geometric, almost mathematical look and is considered beautiful by many, too. The Chrysler building in New York City is arguably beautiful for its incorporation of geometry in its own right; it graces the NY skyline like the cherry on top of a great big milkshake on a hot summer day. A variety of consumer products have been designed using the same Art Deco design language.

FIGURE 28

CHRYSLER BUILDING AND SET OF METAL POTS USING ART DECO STYLE

UGLY AND GROTESQUE

As industrial designers and artists seek to define and understand the origins of beauty, it is instructive to acknowledge the essence of ugly and grotesque. By studying these opposites we can come to a deeper understanding of how to create beauty by defining what it's not. Some of the things that top the list of ugly and grotesque are death, sickness, feces, weakness, and dishonesty. In addition, evil can be very unattractive. Although, some are very attracted to it.

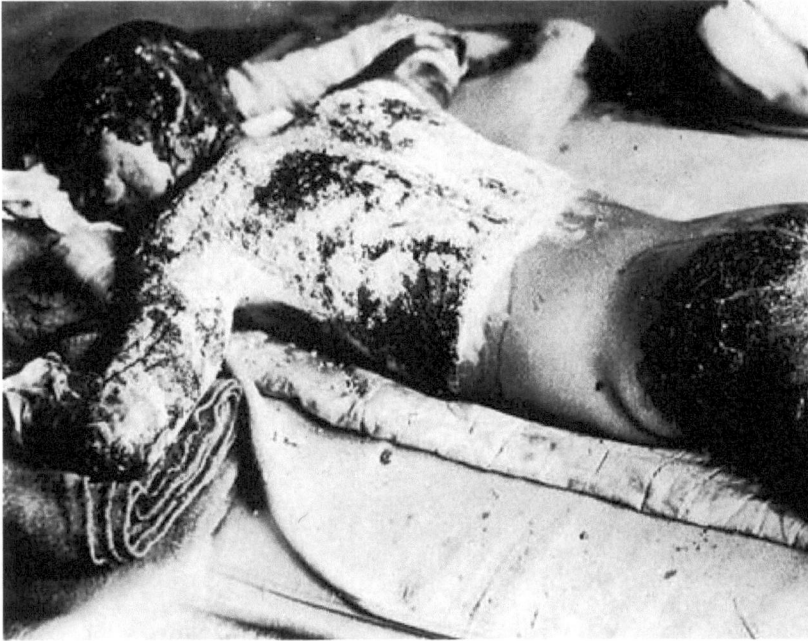

FIGURE 29

BURN VICTIM

Some of the images that go along with ugly and grotesque are blemished, discolored, and uneven features. Things that look wet and frayed look ugly. Things that look decayed and partially done or undone are ugly. The common theme of ugly is damage, dysfunction, and imperfection. Anything that looks dead is a source of repulsion and even fear. On a subconscious basis many of us fear death and want to avoid its inevitable pull. Much of what we do is in an effort to somehow cheat death. Death ultimately takes our control away anyway.

Sometimes death and fearful images are used to create an appealing look, though. It begs the question of why "badass" is appealing and who it appeals to.

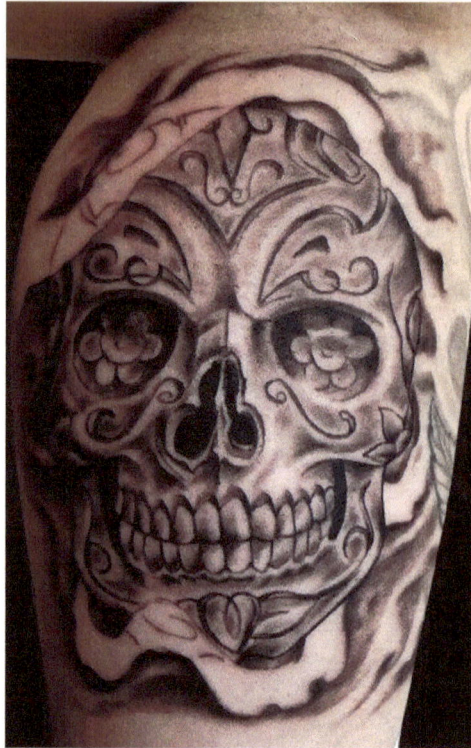

FIGURE 30

DEATH IMAGERY IN A TATTOO

Perhaps the reason these images are appealing is based on a counter culture ethic. Deathly and grotesque images can symbolize freedom and hedonism because they contrast to what is socially considered beautiful. It may symbolize a disconnect from the discipline imposed on us by traditional elements of society. When deathly images are used to create "beauty," in a certain sense death, the ultimate enemy, is toyed with and therefore controlled. If one can make fun of death, or at least take death lightly, even use death to adorn one's body, then on some level death is beaten. Who wouldn't want to be "badder" than the grim reaper?

DISHONESTY IS UGLY

Dishonesty is ugly. When you see simulations of things that aren't done well and unnecessary geometry on things that are meant to function simply, there is an inherent dishonesty that is detected subconsciously if not consciously. No one likes to be lied to. No one likes to be made a fool of. When you see scoops on little underpowered cars, or aftermarket spoilers on cars that would have a tough time pushing 90 mph or hubcaps with simulation spokes on them, it looks cheap and weak. On the flipside, if you create a product that uses authentic materials and, although it sounds cliché, lets the form follow the function, you will have an honest design.

FIGURE 31

AUTOMOTIVE DESIGNS

When a product shows carbon fiber or wood grain and it's obvious that it's really just plastic, it doesn't look good. It looks like a poor attempt to be something else. No one likes a "wannabe."

FIGURE 32

SIMULATION WOOD GRAIN

A lie can look good if it's a good lie though. There are a lot of plastic parts that are coated in a thin metal that appear in many fine products. These parts save weight and cost, but you would never know that they weren't all steel.

FIGURE 33

BMW PLASTIC DOOR HANDLE WITH METAL COATING

THE COMMON THREAD TO ALL BEAUTY

Somewhere in all the many paths to beauty and the opposite paths to ugly there is a common thread. Perhaps in nature, incredibility, extraordinary human effort, relief of human needs, size and balance, there is the inherent representation of the things that are special in life. The fact is most of us desire to be special in some way. Subconsciously there is part of us that knows that if we are just another human we will have to die one day like everyone else. Perhaps for those deep-seated reasons we are attracted to that which promises us relief from human wants and desires and even our human mandated death. Perhaps the products that are based on mathematical shapes are attractive to us because math and geometry are universal and will never die. Maybe the things that repulse us signify death, and we revile them because most of us don't want to be close to it. We don't want to have anything to do with it. Or maybe the things that are beautiful to us are beautiful just because they are. As Freud is credited with saying, "Sometimes a cigar is just a cigar."

DESIGNING FOR THE USER

You must put yourself in the shoes of the people you are designing for to get the best result. Here are some details:

CAPTURING USER NEEDS

From focus groups and surveys to hidden camera studies, good industrial designers stop at nothing to capture the needs of the users of the products that they are designing. User needs, also known as "the voice of the customer," are what should rightfully drive product innovation and evolution. In some industries, there are formal systems in place to document and facilitate the process. In other industries it can be quite haphazard and free flowing. If you've ever seen how companies who design and manufacture strollers operate, you will see a very formal process that begins with a room full of all the strollers on the market. Formal meetings and timetables are established, and the use of a product life cycle management system to capture and rate key requirements and manage them throughout the entire innovation process helps things along.

On the other hand, if you study the evolution of the mountain bike, it becomes obvious that the voice of the customer was expressed through welded gussets and fat tires on previously unmodified bikes. Needs were expressed in terms of the product's failure: in skinned knees, broken spokes, and bent handle bars. Both methods conform to the industry and the people who are involved in the process, and both have their merits.

One of the most important tools is collocation, putting yourself in direct contact with the user experience. "Don't judge a man until you walk a mile in his shoes" is a motto that comes to mind when thinking about collaboration. If you really want to design a better mountain bike, get on one and ride with a group of people who have been doing it for years. Get their stories and invite them into the process. Become the investigative reporter who gets the inside scoop, whatever it takes. With most products this is easy. Every one of us, by virtue of the fact that we all use products every day, is qualified to help in the design process, and we're all critics. In general, people enjoy talking about their favorite products and what they would do to make them even better.

In many cases these people are only accessible when you are in the environment where they are using the product. For example, years ago I had the privilege of helping to design a new generation of poison gas meters. A visit to the local hazmat team was invaluable. As we walked in, many of them had just been weight lifting. They're all constantly prepared for the time when they might have to carry a heavy pack up a hundred flights of stairs. From talking to them we learned about ways that the team used their meters that we never would have understood or considered otherwise. Looking at the thick gloves that they wore when they used the meters, we knew that our buttons had to be larger than usual. Hearing their stories of how they used the meters in a smoke-filled room with low visibility told us that we needed a bright screen with controls that you could feel without having to look at. We learned that they take readings from different heights in a room, which we would never have known without talking to them and having them demonstrate how they would use the product. Epiphanies such as these made a good product even better, ultimately giving us a successful design.

As you discover the needs of your users, there are many ways to document them and rate their importance. One of the easiest effective methods is called a bubble chart. As you talk to members of the user community and they reveal the requirements that they can think of, you may ask them to draw a chart with each requirement surrounded by a circle with a radius large enough to represent the relative importance of each item. Such charts are best done as hand sketches and can be really helpful. Once you have the bubble chart you can measure the radii of all the bubbles and list them in order from largest to smallest, most to least important.

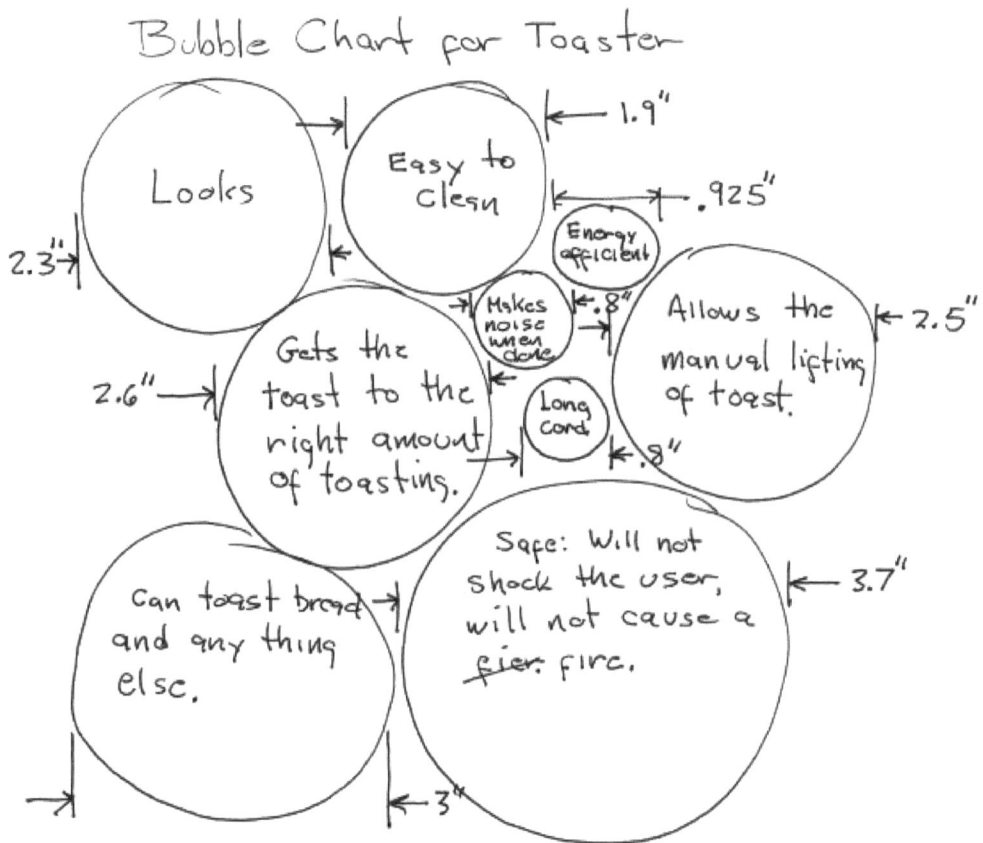

FIGURE 34

Sometimes capturing the voice of the user is difficult because of how varied the users are. There's a product in the gas detection industry called a bump tester that makes sure the gas detection meter works properly. It is used by at least two distinct groups with two sets of needs. A typical user of one group comes around with a soil covered glove and slams a meter into the unit to see if it still has its sensitivity. This type of user is also likely wearing a hardhat and safety goggles. This may inform us on how to design the interface. We may surmise that a highly durable mechanism is needed because of the dirt and the force of a big burly guy slamming the meter into the unit. A typical member of the second group, on the other hand, comes around with a clipboard and clean hands and uses the tester to download data on the entire fleet of meters. The two groups have different needs and lead different lives; our product has to please them both.

You must also realize that with industrial products the user doesn't always make the purchasing decisions, so a product has to work for two different people at different times. It must work for buyers at the time of purchase, and for users through the life of the product, and it has to work for the people who maintain the product and those who retire the product. It should work for the society in terms of its ability to be recycled or disposed of safely and cleanly. All of these sometimes conflicting needs have to be weighed and optimized.

If you're lucky, the voice of the customer is easy to get because they tell you through blogs, chats, tweets, and other types of electronic reviews. Naturally, if there is a way to talk directly to a group of users through some type of club or other group activity it makes the process far easier. The challenge in these situations is making sure that you don't give emphasis to a set of needs that is expressed just because it is expressed better. For example, when designing a bike, perhaps you are talking to a guy who's been doing testing on the front fork. He's got loads of data about shocks and weight distribution and angle of attack. You may be inclined to design your bike around this testing, not realizing that your target market would really be better served by a big, simple beach cruiser.

Sometimes you have to actively give the customer their voice because what you're creating is so different and unusual that there's no one that has any significant experience with your product. Let's say you are creating a tie clip with LED technology inside of it and an accelerometer sensor that changes color when you are moving a lot as opposed to when you are sedentary. You need to tell the customers that they would like the product. Essentially you have to appeal to the human need to be recognized. The voice of the customer is absent because, although most of us crave recognition, as of this writing, there is no established market for LED tie clips.

Recently I was in San Diego, and a few blocks away I saw these ice blue lights hovering in the air. I was drawn to them and wanted to take a look. As I approached I discovered that there was a street vendor that was shooting a small, inexpensive product into the air with a rubber band. The product had thin plastic wings that would fold back when it was shot, then, when it ran out of speed, the wings would angle out and helicopter it slowly to the ground. An LED and battery toward the end of the product hung down and glowed nicely, giving the user several seconds of the show. The voice of the customer doesn't exist before there is an established market, and this simple demonstration was how this vendor created his market. Whoever wants to make their product a great success would do well to help give users a voice by showing the product off in a very public way.

Maybe the product that you are designing is for babies, animals, or others who are challenged in their ability to communicate. A dog toy, for example, is acquired by its owner, but the user is really the dog they buy it for. If dogs don't like it, perhaps because it's made of a material they don't like or it doesn't feel good when chewed, they won't play with it, and your product will have failed to satisfy its user. If the designer has access to a number of dogs through a network of dog owners that regularly meet at a dog park, it will be infinitely easier to observe the dogs at play, observe various toys in different weather conditions or different times of the day, and try out different designs. The voice of the customer may come as a happy bark that brings a smile to the face of their owner.

There are times when the voice of the customer is clearly expressed in cold, hard numbers. Airlines care very much about fuel economy. There is a formula that will tell you exactly how much money will be saved over a given time period as a result of weight reduction. Years ago, as the display industry was changing from heavy cathode ray tubes to flat panel displays, I was hired to retrofit a fleet of planes for a medium sized airline. It was a complete retrofit, and I was not made privy to any of the mechanical design files of the previous geometry. So, as part of the job, I had to remove ceiling panels in a plane, climb up in the space above the overhead bins, take pictures and measurements and come up with a way to anchor my new assembly. As I made my way through myriad wires and cables I could see the so-called "black box" (which is actually a bright orange-red) positioned over one of the doorways. It was a really neat experience.

The enclosure that I designed, along with the corresponding internal support frame, was beautiful, but in the end that is not what mattered to the customer. They were really only focused on the bottom line weight reduction number. The design I was able to implement was about a 60-pound reduction per screen, with five screens per plane. The customer was pleased to be saving hundreds of thousands of dollars each year due to fuel savings; the beauty of my design was definitely a secondary consideration.

Some user needs, like those of airlines, are a lot easier to capture, quantify, and organize into design requirements than others. Physical properties such as speed, weight, height, and temperature can be easily handled. Other aspects of product design are best done with samples. For example, if a product is covered in cloth, and the product has to be a color that goes with the current fall fashions, it's difficult to quantify and accurately describe how to coordinate those colors. Since different colors go in and out of style with the latest fashions, one has to examine many different shades used in different industries and the color related history of the products that are "doing well" as opposed to those that aren't.

There are software packages specifically written to formally capture the requirements expressed by user communities. For example, Teamcenter Requirements Management created by Siemens is a package that helps designers and inventors assign identification tags to lists of formal requirements that are associated with a new product. As the design process continues, the requirements are tracked. This guarantees that the design meets every requirement. This formal approach is incredibly important to the aerospace industry and the medical industry where products are highly regulated by nationally recognized testing labs, and the outcome of a requirement that isn't met can be a matter of life and death. During the design of certain classes of medical products manufacturers are also required to produce a design history file and many other formal documents. The requirements of these types of products can be extremely technical and detailed. There can be material requirements, performance requirements, heat loads, sound emissions, and many other attributes.

THINK OF THE PERSON, NOT THE THING

A great product designer thinks about the person that the design will serve first, and then the actual product after that. For example, when you design a chair, it's important to know who the chair is for and what purpose the chair serves. For example, the queen of a tiny island nation wants a chair for her to meet dignitaries in. If you're not imagining the experience that she wants when she meets the dignitaries, you will miss the point. You may focus on comfort and not realize that the purpose of the chair is to make her look grand, perhaps grander than she really is. The chair must elevate her, and it must send a message to the dignitaries that she has wealth and power. Perhaps the chair should be covered with jewels—not because they are beautiful, but because they're hard to obtain. Perhaps the chair should be elevated so that the queen has to climb up into it. This would make her seem more powerful and allow her to look down on the dignitaries. It's not a chair that I would like to design, but if that's the person who you are designing for, then that's what your design has to do.

Great product design means the product works well for the end user. You as the designer should know what the end user of the product does with your product. For example, I recently purchased an inexpensive table from an office supply store. It was for a booth that we had at a trade show. What I wanted from the table was to be able to set it up quickly to put flyers on. It was made out of plastic with inexpensive steel tubing underneath. When I went to set it up, I took it out of the box and I had to take off the wrapping. Someone at the company who made the table got the bright idea of wrapping the plastic table and then attaching the leg assembly down over the plastic. I'm sure they saved about 15 cents worth of plastic and twice that much in time, but the end user, myself, was pissed off. It was impossible to get all the plastic off without dismantling the table leg assembly. Even now, there are little bits of plastic stuck on the underside of the table in between the leg and the table top.

The designers never thought of the user experience. All they were thinking about was the price. Now, I'm fully aware that the price is a very important part of the user experience of a product, but having your customer laboriously pick plastic out from your product is not a good trade off.

Let us also remember that the packaging of a product is an important part of the product experience. We all know that first impressions can be lasting. If you let your packaging people put a sticky label on your product so that when the buyer gets home with the product they have to use their fingernails to get the label off and they get that annoying residue on it that they just can't seem to get off, *you* have failed as a product designer. Many professional design firms completely miss this point. Many products come in packages that piss off the consumer. Every one of us has experienced opening a package that is unclear how to open or is difficult to open. Most of us have also experienced packages that were so tight that we had to use a tool to open the package. The package has to convey a marketing message, but it should be designed well, too. The package should do more than influence a purchase decision. It should be easy to open, protect the product, and be easy to stack on top of each other. A great package should also be biodegradable.

Like many other aspects of product design, you will probably do very well by finding a good packaging manufacturing house. Choose a packaging company that has an office where the product is made, that has access to the latest biodegradable materials, and that knows how to do the analysis required to ensure that your product will make it safely to the consumer.

LET THE EXPERIENCE CREATE THE FORM

Here's another great method that you can use to create a great product. It's another cliché: Let the product be designed by the experience. For example, let's say that you are designing a new toaster. Begin by imagining that you want toast in either the easiest or the coolest way. Close your eyes, take a piece of un-toasted bread, and load it in something that will toast it. What would make you the happiest? In a perfect world you would want the toast to be ready for you perfectly toasted to the level you want. You want it to be even, and you want it to call you when it's ready. You don't want to have to clean it out and you never want the bread to get stuck in the toaster. With your eyes closed, you can imagine a toaster that you place a piece of bread in or on, and a toasting head that moves over the bread like a fax machine. Perhaps you can program the head to toast various patterns into the bread with lighter and darker patches. Perhaps you can get the bread toasted faster with less energy than a normal toaster because the toaster doesn't have to toast the entire thing all at once, and it doesn't have to heat the entire cavity.

We've all had toasters that aren't big enough to take four pieces of bread; maybe you can produce a new toaster that can toast as much bread as you can stack in it because it only toasts a portion and then moves on. As you can see, visualizing the experience of getting toast will lead you to new and often better solutions than if you start with a concept: start with the solution rather than the problem.

Just recently, I was driving a rental car and I looked down at the instrument panel to turn on the heat. To my shock, I noticed that some of the labeling was under one of the knobs. I couldn't see what it said because the position of my head was high above the instrument panel. It was obvious to me that the designer of the dashboard designed the graphics looking straight forward on a large computer screen. That designer was designing a dashboard, he wasn't designing the experience of using the controls on the dashboard. In a way that designer failed to care about the user. In order to design a good dashboard, you have to imagine yourself in the driver's seat and think about the experience you want. You shouldn't have to look down and underneath a knob as you are driving —especially since many of us drive above the speed limit.

In one of my favorite cars—I won't rat it out—the heat is controlled by two buttons, one with an up arrow and one with a down arrow. This is ludicrous. When the heat is all the way up and you want to turn it off you have to keep hitting the down button many times. You can't just turn the heat off with one touch. Also, the same car has a wheel for the heat that comes from the vents in the middle of the dashboard. Not only are there two different controls, but the wheel has a graphic on it that is painted on and not backlit. This means if you are driving at night and you are not yet familiar with the car, you can't see how to turn the heat off. It's clear that the dashboard designer failed to see the experience of using the dashboard from the perspective of a new user. In this day and age, that's surprising.

LEARN CONTINUOUSLY

It is very important especially in the design field to be a life time student. Here are some details:

LEARN FROM TEARDOWNS

A teardown is a wonderfully destructive process in which you obtain a product that you admire for whatever reason, and you take it apart. Many products are glued together or sonic welded so that when you take them apart you have to break them. Not to worry, you will learn so much that it will be well worth it. You don't have to tear apart a new product; you can get it used from eBay or a yard sale. A teardown is one of the best ways to learn about new design configurations.

I've personally been extremely fortunate in this respect. When I was a young boy in the sixties, my father, an extremely creative and patient person, would take my older brother and me to a junkyard. In those days, you could just wander onto the property and dig around for whatever you could find. We would find old pieces of anything and everything, gather the most interesting ones, and take them home. My mom wasn't too happy about it, but she supported my dad's efforts to teach us kids. She was an incredibly hard worker, and she really knew the value of being strong and learning, even if it made a mess.

We lived in a three-bedroom apartment in Brooklyn, NY. We didn't have a garage space so we went out on the terrace. We took the junk apart and in so doing we saw how things were put together. In our household the ethic was to fix things. If the toaster stopped working we would get out the screw drivers and the pliers and see what we could do. When the TV stopped working we would open it up in the back and see if there were tubes that seemed to be burned out. Our first vacuum cleaner was an old, discarded Electrolux model. We took it apart, fixed what was wrong, and it served our family for years.

In many ways, the old days were more mechanical. You could fix anything because things were not built with as many complex or computerized parts and the manufacturing techniques were far simpler. In these days of automatic assembly and numerically controlled manufacturing techniques, products are a lot more complex. Back in the day anything you needed to do on any car was within your grasp. These days a lot of products rely more on integrated circuits. When they stop working it's more of a mystery. Even so, there is still a tremendous amount to learn from teardowns.

As you tear down products it's a good idea to have a camera and a pair of calipers on hand and to be able to note key material thicknesses, screw sizes, hole sizes, and more. Go ahead and take pictures of various mechanical configurations and make a scrapbook of really cool solutions. It's very similar to the way a good writer will copy down key sentences from works they are reading that they really admire. These notes will help you remember the powerful mechanical solutions that make up the best products. The knowledge that you gain from teardowns stops you from reinventing the wheel, letting you harness the power of what's already out there.

LET THE SUPPLIERS EDUCATE YOU

When you're trying to come up with a new design, you are at a huge disadvantage if you don't know how things are manufactured. You should really know about plastic components, machined components, carbon fiber, sheet metal, wood, etc. You should know about welding, fastening, adhesives, and other manufacturing techniques. You can spend a lifetime taking courses at a major university where you will be taught the academic version of the latest techniques for manufacturing. However, you may not have time to attend university while you are trying to make a living. As an alternative, sometimes the best thing you can do is talk to a manufacturing supplier. In my experience they are usually happy to give you a short plant tour and let you know how to change your design so that it will be less expensive to manufacture.

Let's say for example you are designing something in sheet metal. A single call to any sheet metal supplier anywhere should get you the facts that you need, such as standard available sheet metal thickness, minimum bend radius, the properties of various metals, etc. Suppliers are not generally design experts, but they sure know a heck of a lot about their niche. Recently here at Design Visionaries, we had a representative from a company that does electroplating come and visit us. He was happy to bring samples of his work, and he allowed us to ask all kinds of questions. He even let us keep some of the samples so when we were designing we could think of him and the processes that he taught us about. People who manufacture stuff are constantly amazed at the naiveté of the folks who submit designs. Especially these days with any modern CAD system, it's easier than ever to design things that are impossible to make with standard high-speed manufacturing techniques. Just because you can bang out a prototype on a rapid prototype machine doesn't mean you have a good design that can be manufactured cheaply. At the time of this writing rapid prototyping was still not to the state where it is widely used for production parts, though that is slowly changing.

LET YOUR INDUSTRIAL DESIGN BE INFORMED BY OTHER PRODUCTS

The look of a product can sometimes be as important as the actual function. The problem is how to make a product look "good" when good is so subjective. There are a few known examples of quantifiable good looks in the industrial design world. The golden ratio of 1 to 1.618, which is the ratio of the side lengths of the "golden rectangle" and the G2 continuity of curvature, which ensures that the light from a product is soft are a few common ones, but examples like these are few and far between.

When considering how to approach the look of a product, again consider the user. Think back to the principle from the beginning of this section: "form follows emotion." Emotions can be a difficult thing to capture and convey for your user, especially when your life experiences differ significantly from theirs. Various groups of product buyers and users have emotional connections that are very hard to anticipate. For example, if you are a design type raised in a big city, you may have very little connection and knowledge about the midwestern cowboy who goes hunting. When someone asks you to design a device that works for the cowboy, it's difficult because you may have to design using someone else's voice. You may immerse yourself in cowboy culture, you can visit, go hunting, attend a huge barbeque, but you may still feel very little connection to the group of people that you are serving.

One great way to come up with features to make a product look good is to gather up a number of "good-looking" products and make a collage. The products that you gather can be other successful products that the target market uses. For example, if you are designing a product that will be sold to participants at a monster truck rally, you may decide to gather images from a number of related magazines. You might go to an auto parts store and gather a number of tools that have been well designed and then go to a sporting goods store to see what's there. These products can be your guide. Once you have your collage, you can begin to borrow visual cues from those products. You may be able to borrow surface textures, colors, shapes, materials, and many other aspects from them that you and your users will find appealing. This is another aspect in which user feedback is powerful. You can gain a wealth of knowledge from people who use similar products to see what they like about its appearance.

PRACTICAL DETAILS

This section contains other things you should know that may help you to avoid a million-dollar mistake.

LET THE FORM BE FRIENDLY TO MANUFACTURING AND ASSEMBLY

Professional industrial designers do it all the time. They go off on these ultra-cool design excursions and they produce beautiful drawings of products that are impossible to build. They love it. They're trying to win design awards and some of them don't have the slightest care about the rest of the design team or even the user. When the product designers, the mechanical folks, get the design rendering from some industrial designers and are trying to figure out how to make it a reality, they often scratch their heads. "How does he expect us to fasten these parts?" or "How are we supposed to injection mold these parts when there's this huge undercut?" or "How does the designer expect us to fit the motor into this design when the neck is that small?"

Problems like these can cost millions. They delay the project with endless back and forth communication, and sometimes lead to bickering between the industrial design group and the product design group of major corporations. When the industrial design of a product is done well, it helps the rest of the design group. The product designers find that the shapes are beautiful but have been carefully considered so they can actually be manufactured. The manufacturing folks find that they can actually produce the required bends in the sheet metal and that when they begin to define the tooling it's actually easy to do so. They don't need to invent some crazy new process just to achieve some strange looking design. Let there be no mistake, sometimes shapes are required that need a brand new and creative manufacturing process. Some of the most successful and beautiful products owe their successes in large measure to new looks that were achieved by a great design and a brand-new manufacturing process, but these are the exception not the rule and should only be required when absolutely necessary.

Similarly, many types of products need to be assembled. From a two-piece plastic toy to a complicated MP3 player, the assembly of a product is a very important consideration. If you have done a good job in the mechanical design phase, a lot of the assembly considerations have been taken into account in the engineering phase. There is a lot to know about the various techniques of assembling components. Much of it can be learned by taking existing products apart. For plastic parts, there are various techniques of assembly, each of which has many pros and cons. For example, self-tapping screws are good because you don't have to create threads in mating parts, which saves money and time. There's "heat staking," which is the method of leaving a small boss in a part that fits through a hole in a mating part before an iron is used to melt the boss in the hole. Heat stakes are good because you don't have to manage another part number such as a mechanical fastener. Reducing part counts is usually a desirable goal.

Another technique is called sonic welding; it's the process through which two parts that must be joined are vibrated together, which causes melting where they touch. It's very desirable because it's very fast and secure; however, special equipment is necessary. Using snaps is also very desirable because they usually allow quick assembly; their downside is that they usually complicate the molding of a component due to the creation of "undercut" geometry.

Perhaps one of the most undesirable methods of assembly is using adhesives. Adhesives can take a long time to fully cure; they can be messy and difficult to apply. Sometimes they must be done in controlled environments, they can be a health risk, and they can require special and expensive equipment to dispense. Sometimes, unfortunately, they are the only possible method. Industrial strength double sided sticky tape is a common form of adhesive that is used.

CREATE A WELL-THOUGHT-OUT PRODUCT REQUIREMENTS DOCUMENT

In my experience, creative people like to go for it. They're not so apt to take notes, write things down, and worry about formalities and details. In many cases this is a huge blessing because they get off their butts and do something. They passionately chase after a vision and in many cases throw caution to the wind and create something that no one has ever seen before. So creating what is known as a Product Requirements Document (PRD) is not something that always comes naturally even though it can be extremely helpful—especially when there's more than one person involved in an endeavor.

A typical PRD lists out many of the things that the user of the product will experience with it down to the minute details. This is essential information for you as a designer as well as for anyone you end up working with. A good PRD will include information like how many buttons the product has, how many LEDs it will have, what the buzzer will sound like, how long the product will wait before it switches modes, how many sensors it will have, whether the product will be battery operated or if it will need to be plugged in, and other details that pertain to your design in particular. It will have some information regarding the user as well. For example, if the product is meant for children, it will probably need to be battery operated and the buttons may need to be different from a product made for adults.

The PRD ensures that there is a clear goal for the product that will be created and helps to organize all that has to happen as the product progresses. The PRD also contains the answers to many of the questions that are most important to the success of the product. For example, "Who's the product for?," "How much is it supposed to cost?," "What are the features and functions?," and "Where and how will the product be used?" As these questions and many more are systematically written down and answered, the developer is forced to contemplate. As they contemplate, the definition of the product begins to emerge. As each design attribute is addressed the product takes on a more complete definition.

For example, an inventor has a novel idea for a toaster. It's not an industrial toaster; it's for the home market. Right away there are implications that have been set forth for the product and all the electronic goodies that are inside. The product has to do well with four large slices of bread, and bagels, too. Now we know a bit more about the product. The target user is upper-middle class. We now know that a large footprint may be acceptable. The toaster is supposed to be better than most at dialing in on the amount of toasting that occurs. Now we know that we may have to have some advanced technology packaged in like a light sensor or some kind of infrared temperature probe. Based on our knowledge of the user we can determine if the price point will support any kind of advanced technology. The PRD may define the sound that the toaster makes when toast is ready or it may specify the light intensity and color of the little LED that tells you when the toaster is in use. Typically it should cover every dimension of the product: its weight, look, function, cost, packaging, recycling method, smell, and everything else conceivable.

The PRD also helps to avoid or at least keep track of project scope creep, which is when a product's requirements keep changing and growing continuously. Project scope creep is often necessary but is almost invariably disruptive and costly. It negatively impacts the schedule, in many cases requiring you to go all the way back to the industrial design phase to make changes to the product that reverberate through the entire project definition. Scope creep sometimes requires more research and can be exorbitantly expensive when you have to redo major portions of the design and manufacturing planning, and it can make you miss selling opportunities in a competitive market place.

Finally, the PRD is a communication tool. Design, marketing, manufacturing, maintenance, and any other of the cross functional groups that make up most companies can all agree on what is to be innovated, what their role is, and, perhaps most importantly, how their piece of the product coordinates with all the others.

CRITICIZING YOUR WORK

Whenever you are designing something new, you may go through a lot of ideas. Some of them will be good and some of them will be pretty lousy. The problem is it's not always easy to know which is which. It is a good idea to write down or record these ideas without being that picky right away, but at some point you have to pare it down to the few that you will invest real time in. In many cases your ability to criticize your own ideas is directly proportional to the amount of energy you have at that moment. When you have very little energy and you look at the work that you've done you know deep down how much effort it's going to take to fix it. You can sense the hours or days it may take to do something else, and you have already invested a lot of time. You don't want to lose that time. When you have worked long and hard on something and you are at the end of your patience, it's easy to say "Ah, that's good enough, no one will ever notice such and such." But they will. If you give yourself time to step away from a design, refresh, and come back later, you may suddenly have the patience to look it over more clearly. You will see flaws in your work that you didn't see before, and you will actually have the energy to fix them. In many cases when you set out to fix them the thing that you thought was going to take you so long and be so arduous takes you far less time than you thought it would. Sometimes you can take an idea that is pretty bad and add to it in such a way that it becomes a good idea in the end. Self-criticism and rework serve as the great improvers of whatever you may do.

Even then, our own criticism can have its limits. Don't bet the farm on an idea without talking to industry experts and doing your homework. We inventors are full of passion—sometimes to a blinding degree. It's like when you write a term paper. When you proofread it, you may not realize that one of your sentences is utterly unintelligible because you know what you're trying to say and unintentionally overlook the mistakes. The industry expert can be an objective viewer. The right industry expert will be critical and may have a chance of seeing something that you missed. We have all heard of stories where a person had a great idea that no one else could see or appreciate. They invested everything they had, went broke, and then finally prevailed. This is the exception, not the rule. It might make a good story, but you probably don't want to set out for that kind of experience. Especially when you're starting out, catastrophic failures may not be an option. There are many successful inventors you've never heard of that do their homework and have many inventions. Successful inventors create good plans and take care of the details. They get funding, and they minimize risk. If they fail, they live to fight another day.

Improve your idea. Every invention that has ever been created can be improved. This is a good attitude to have about inventions and products alike. Even the wheel has been improved by the advent of lighter materials, aerodynamics, and then magnetic levitation. If you always try to improve upon your initial idea, you will have a better chance of producing a product that is as good as the current technology allows.

Having what you think is a great idea is exciting. You will probably want to share it with family and friends—how can you resist? However, keep in mind that they may not necessarily be objective. In an effort to be supportive, some friends and family may be too positive about an idea. They would love to see you progress. On the other hand, some family, and, more often, some friends, have a vested interest in keeping you in the exact same position you are in. Deep down, they may not want to see you progress because it will make them feel left behind. This may skew their view of your idea. Some of your friends may have tried and failed before, and, although they mean well, they may be unnecessarily harsh on your idea. The bottom line is that there is an emotional involvement, and you have to take what they say with a grain of salt.

COMMUNICATING YOUR IDEAS

Communicating your design idea is extremely important. Ultimately nothing can progress if you don't document your idea in a way that your manufacturing partners and all other parties can understand and act upon. To follow are some basic yet powerful techniques:

HAND SKETCHING

The best ideas generally come from a vision of how things can be improved. In the case of product design, that vision may be augmented by a designer's 3D imagination. It is a great advantage if you can vividly picture the product or component in your head. Once that vision is somewhat clear, making a hand sketch should be the very next step. For any idea that involves mechanisms, even if it's a fancy electronic device that needs a housing, please, please, please, do hand sketches first. Many inventors, especially young ones, tend to jump right into computer aided design and work long and hard to create accurate product definition data, only to find that there is some fundamental problem they could have easily seen with some hastily drawn hand sketches. Creating a hand sketch forces you to visualize an idea in a way that CAD does not. A hand sketch enables you to hone that imaginative mental model into something that can be seen more clearly and easily improved upon. It can play an incredible role as far as digging into the details that make an idea really work.

In the same way a woodworker may begin with a rasp and work her way down to coarse sandpaper then successively finer grain sandpaper, the designer can work a series of hand sketches, each one getting successively more clear and detailed. If space allows, a very useful technique is to have successive sketches splayed out on a large table so you can easily borrow qualities and features from one sketch for the next. You may use tracing paper to create sketches on top of previous sketches then add more detail. You might even use a photograph as a backdrop to sketch a device that may interact with another device on the photo.

Although they are helpful, hand sketches don't have to be drawn in perspective or even be geometrically proportional. The sketches don't have to be beautiful or even "good;" they serve their purpose regardless. Even a quick sketch on a napkin drawn with lipstick can give the designer a huge advantage. More detailed sketches are, of course, incredibly helpful as well. The more advanced sketcher will use effects such as shading to capture the curvature of proposed products and show complex reflections, but that is not always necessary. Other techniques such as section views, detailed views, and broken views can give great insight. Even for complete novices, it is surprising how quickly you can learn and use these techniques with just a bit of practice

.

A QUICK GUIDE TO HAND SKETCH TECHNIQUE

Here's a brief overview of basic techniques and features to keep in mind when creating a hand sketch.

GUIDE LINES

In many cases, you can begin a hand sketch by lightly sketching simple shapes with a pencil before you sketch in the actual product with heavier lines. These light shapes can be used as rough guides for the details later. The light lines that you begin with do not have to be erased once the sketch is done. In fact, they may be more useful to keep around because they can provide helpful visual cues.

SHADING

Shading can be added with just about any writing utensil you choose; all you need to do is draw some lines. To add shading to your drawing, first imagine where the light is coming from. Note that the things that are closer to the light will appear lighter than the more distant, shadowed areas where the light is blocked. Now, using your writing implement of choice, make a number of lines on those shadowed areas to represent the darker appearance. Make more lines where there should be more shadow. For a cylindrical figure, the shadow is easily represented with a number of lines that are closer as they get to the ends and not as close where they are in the light as shown in the following drawings.

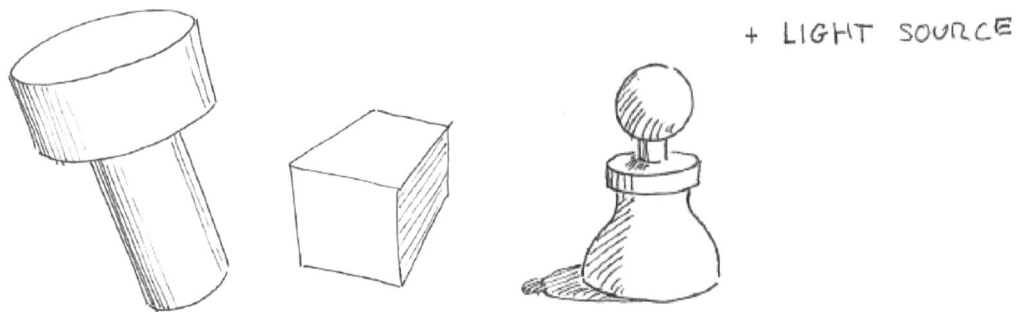

FIGURE 35

THREADS

It is tedious and usually not important to draw accurate individual threads on a product sketch. You don't need to make them exact or evenly spaced in your sketch. A quick and easy way to represent them is by drawing a cylinder with zig zag lines where you would usually draw the straight lines of the sides. Then, make a number of slanted lines to show the threads.

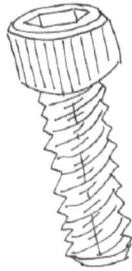

FIGURE 36

VANISHING POINT AND PERSPECTIVE

It's a pretty basic principle; to the human eye, things that are far away appear smaller than things that are close. Imagine you're standing in a hotel hallway that stretches on forever. The corners of the hallway seem to recede until they reach a single point. The doors, weird bland artwork, and even the pattern on the carpet shrink similarly, angled toward the vanishing point. On paper, the same principle applies. When we want to make something look far away we have to draw it smaller than what is close. When drawing a box, imagine it fitting snugly inside that receding hallway. The edges of the box will be angled toward that vanishing point, giving the appearance of a 3D object. When you draw with respect to a vanishing point it will make your hand sketches really pop and give them a sense of realistic perspective.

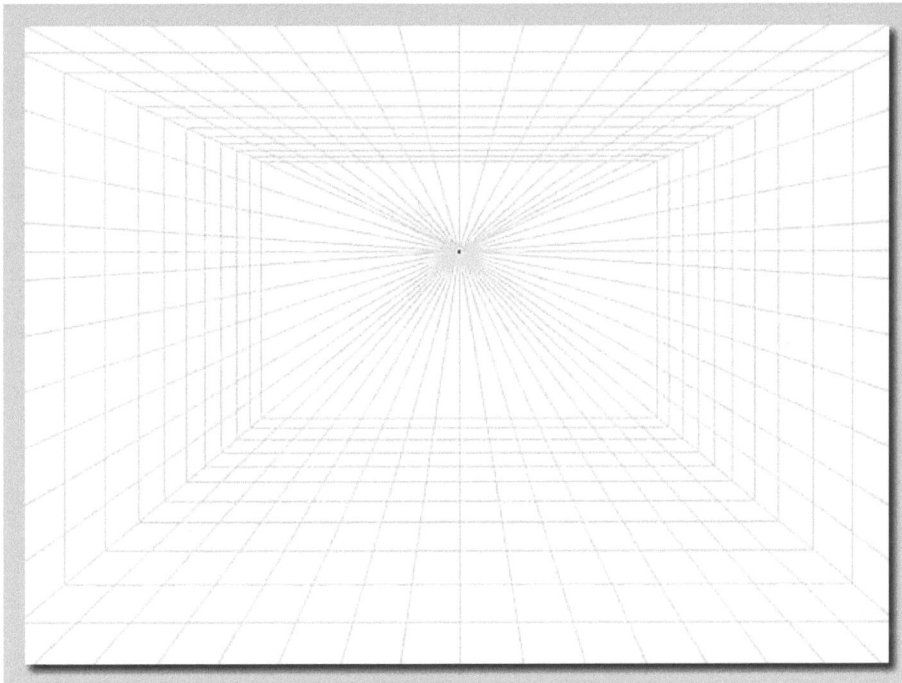

FIGURE 37

ORTHOGRAPHIC PROJECTION

Imagine you are holding a tiny version of an American minivan suspended in the center of a glass cube. First position the cube so you are looking through the front face at the grill and windshield head on. Then, rotate the cube 90 degrees so you're looking at the driver's side profile of the car. You have just performed the rudimentary skill needed for an orthographic projection. An orthographic projection allows you to represent a 3D object in two dimensions by providing multiple views of that object.

To translate this concept into a sketch, imagine you then project the lines you can now see on the driver's side onto the glass face you're looking through before unfolding that face so that it lies flat and in the same plane as the original front view. That gives you the right view of the car. Performing the same process for the other sides will give you the top, left, and bottom views.

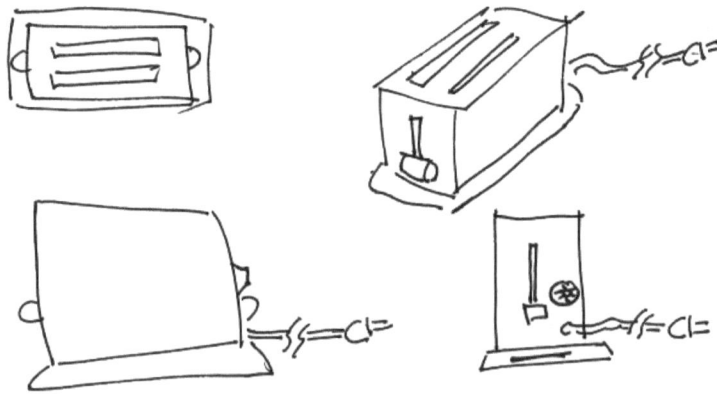

FIGURE 38

SECTION VIEW

A section view is the result of cutting away a piece of a 3D model so that you can show what's inside. When you draw a section view, draw a line to indicate where you would make a theoretical cut in the subject that you are sketching. The line has arrows that point one way or the other and the view you've created is placed on the side of the section opposite to the direction of the arrows.

BREAK OUT SECTION VIEW

The break out section view is like a miniature section view that doesn't go all the way through the part. It's a localized section that shows the detail of some small feature that probably doesn't go the entire length of the part.

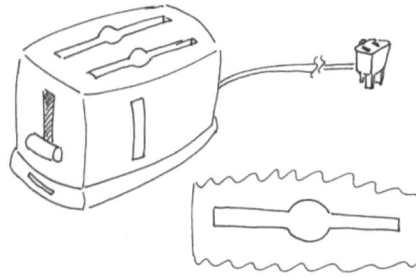

FIGURE 39

DETAIL VIEWS

A detail view is like a magnifying glass that is used to show the details of something that would usually be very small and hard to see in the context of the overall sketch. For example, if you wanted to sketch a door and you wanted to show the detailed shape of the keyhole, you'd make a small circle around the keyhole feature, then draw a larger circle off to the side with a zoomed-in, detailed sketch of the keyhole inside of it. You may label the small circle with a letter, then label the larger circular view with the same letter. This way you can show multiple detail views in the same sketch without getting them confused.

FIGURE 40

EXPLODED VIEW

The exploded view in a hand sketch shows the various components of a product expanded outward as if they were all moved away from each other in straight lines. This kind of view lets you see what's inside the product and where the various parts fit. The exploded view doesn't always have to show every single component, but it can be created such that even the smallest fasteners are shown if necessary. The exploded view can also be accompanied by a bill of materials, which is a list of all the components and their quantities.

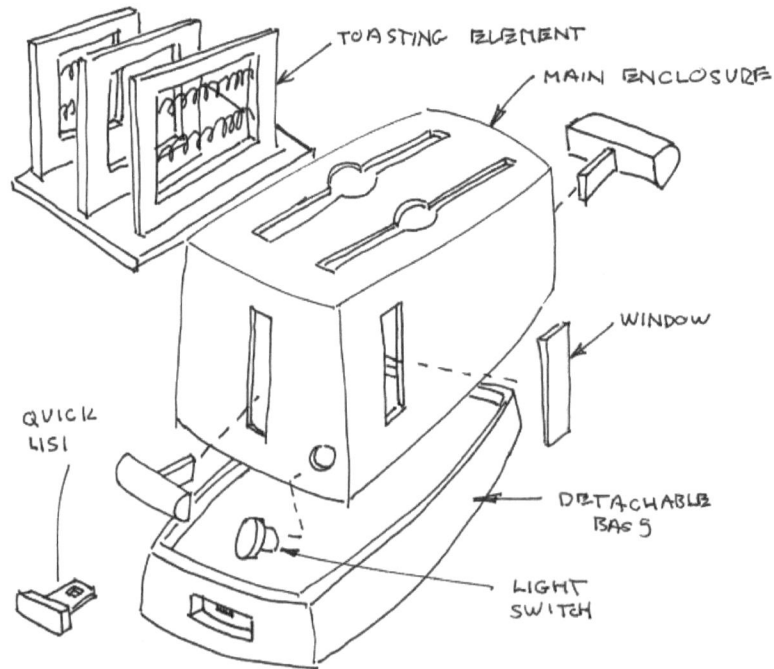

FIGURE 41

MAKING USE OF HAND SKETCHES

The ability to hand sketch can make a huge difference when you are presenting an idea to a prospective collaborator, or even someone who you may want to have help out with funding. That hand sketch can communicate a complex idea while giving them confidence in your abilities and knowledge. It is exceptionally useful when presenting to those who may have very little 3D imagination but may be of great help in managing, financing, or marketing. In these cases it can be useful to perform high quality hand sketches. There are a number of additional techniques that make this possible and form the basis for high quality product renderings.

A wide variety of skills are required to produce high quality hand sketches. It sure helps to have copious amounts of natural talent and it may take years of practice to get really good at them. If you're interested in developing your skills so you can make these high-quality sketches and product renderings, I would highly recommend you check out a textbook entitled *Sketching: Drawing Techniques for Product Designers* by Koos Eissen and Roselien Steur. It's the best book I've ever encountered on the subject.

PROTOTYPE

In most cases an idea can't really be validated without a prototype or some type of manifestation. With modern inventions there is, in most cases, a need for more than just one type of a prototype. You may start with a cheap cardboard replica just so you can hold the invention in your hand or see it work in some way. Then you might build a scale replica out of balsa wood or clay. The trick is to find out as much as you can as early as you can and for as little cost as you can.

At first, get the cheapest, fastest prototypes possible. If you are designing a new kind of linkage for an auto jack, or the lid on a new bucket loader device, by all means, make a scale model out of cardboard before you begin to spend the big bucks. You can learn an incalculable amount from a five-dollar scale prototype. If you need more detail though, there are all sorts of fast and cheap methods these days of obtaining a professional-looking production prototype such as stereolithography, selective laser sintering, 3D printing, soft tool molding, and myriad other techniques. Many of these must be driven by a CAD model, but others are not. Prototypes can be used just for the looks of a product, or to see if some portion of the internal mechanics will work. Prototyping early can easily stop you from spending a million dollars on development of a concept that you will only find out doesn't work. In many cases you can learn more about the behavior of a complex system with an ultra-cheap scale model, than if you paid 10 PhD graduates to perform five months of calculations on it.

SOLID MODELING

Solid modeling is one of the most powerful techniques that we have available to communicate the details of a great idea. Here are some details:

INTRODUCTION

All design starts in the imagination of a person with a need. If water needs to be transported, music needs to be portable, a bridge needs to be constructed, or anything else, there is a great design waiting to be discovered. The design process begins as soon as a capable person says there needs to be a product. If that product has any type of physical manifestation like sheet metal panels, a molded composite housing, or even a wooden frame, the design becomes complex enough for a modern computer aided design system to help. The more complex the design, the more advantageous a CAD system is. For the design of a large jet engine that contains hundreds of thousands of separate components, a CAD system is an absolute necessity. Even though it is possible to create a fantastic jet engine using little more than slide rules and drafting tables, to be competitive in a sophisticated market place high end CAD is essential. For simpler designs (i.e. the design of a wooden chair) the CAD system can still deliver huge benefits to the designer/ engineer who is proficient in it. Solid modeling describes the 3D geometric models that are produced with a CAD tool. Product Definition Data (PDD) is the acronym used to describe not only the 3D geometry, but everything else that goes along with a product definition. This includes material properties, color, manufacturing process, etc.

Perhaps some time in the future, there will be a CAD system that will somehow work on brain waves and render a design to fit the desired application instantly. The design will be perfect in every conceivable way and will automatically translate into the instructions to run the most sophisticated manufacturing tools. The proverbial button will be pressed, and the fully functioning product will emerge. The evolution of CAD systems is a consistent march to that lofty goal. However as of this writing CAD systems and solid models have to be driven using keyboards, mice, valuator devices, and several commands and menus. The most powerful CAD systems have to be studied for quite some time for users to work on at an expert level.

Those that use CAD on a professional level and are involved in design processes over many industries have converged on certain agreed upon terms; MCAD is Mechanical CAD—the term is necessary to differentiate it from Electrical CAD (ECAD), which is used to layout circuit boards. "Solid modeling" is the act of creating 3D models of components and systems as opposed to 2D layout work. An assembly is a collection of those components, and a component is built up of certain entities whose terminology varies a bit among the CAD systems on the market place. Most CAD systems offer some way of creating basic geometric building blocks in 3D space called primitives. These are the cube, sphere, cone, cylinder etc. Other categories of 3D entities are sheets or surfaces. Both terms refer to 3D but volume-less shapes. A user can create flat surfaces, cylindrical surfaces, airfoil like surfaces, and many others that can be converted later into solid shapes. Other important entities are curves such as lines, arcs, and splines. Additionally there are conics, text entities, planes, axes, and coordinate systems. A user can also create helixes, parabolas, hyperbolas, and any other function curves that can be described mathematically.

Once a good solid model is created to represent a part there are a huge number of functions that it will support. You can find out its weight, you can see if it will fit into an assembly, you can create 2D drawings for it or machine cutter paths. You can animate it as it is driven kinematically along a pre-determined path, and you can get all the needed views of it with shading and reflections in perspective or not; the sky's the limit.

SOLID MODELS

One of the most important principles of solid modeling is the concept of the closed and unified volume. A solid model is only a solid model when there is a fully closed volume of surface entities that have the extra information of which sides of the surface are inside or outside. Those surfaces must be unified, and have an assigned density. For example, one may place six square surfaces together such that the shape formed is a perfect cube, but if the surfaces are not unified or associated with each other, they don't form a solid. The example below shows two sets of surfaces arranged in cubes. When a hole is drilled in the one that is not a solid, it opens up to the interior. When the same hole is drilled into the solid, new surfaces are created to represent the material inside.

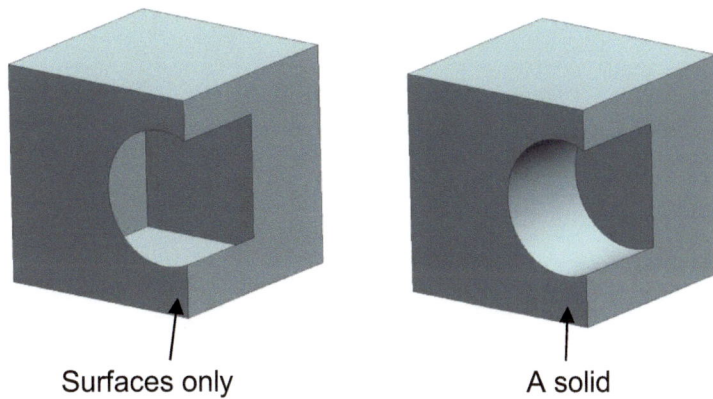

Surfaces only   A solid

FIGURE 42

The modern designer needs to be able to quickly operate on these entities, trimming them, scaling them, copying and moving them if desired. Then needed solids can be added and subtracted from each other to form more complex solid models. For example, when someone subtracts a small cylinder from a wider one, a ring results as shown in Figure 2. Similarly if the two shapes are added a wheel with an axel is produced.

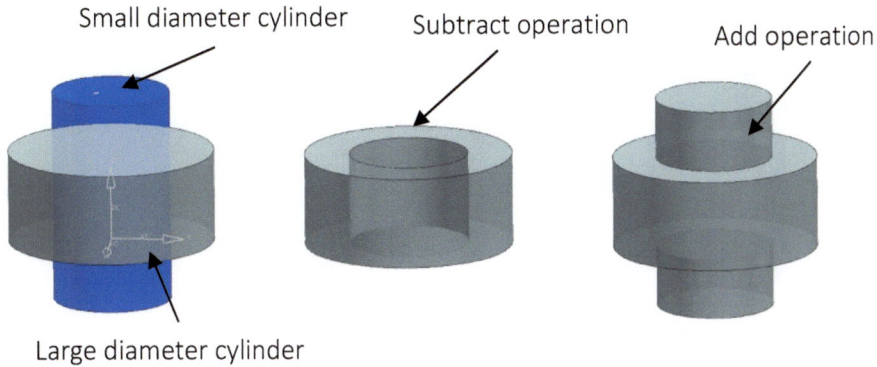

Small diameter cylinder   Subtract operation   Add operation

Large diameter cylinder

FIGURE 43

The essence of solid modeling can be captured by two basic analogies. You may start with a block of material and keep whittling away at it like a marble sculptor, or you may begin with a small solid and keep adding to it like a clay sculptor. You may also use a combination of both processes. For example, you could start with a block as shown in Figure 3. Then make angled cuts on all the sides, scoop out the center, add small cylinders at each corner, and smooth out the corners to make a serving tray. This method is much like the sculptor's approach.

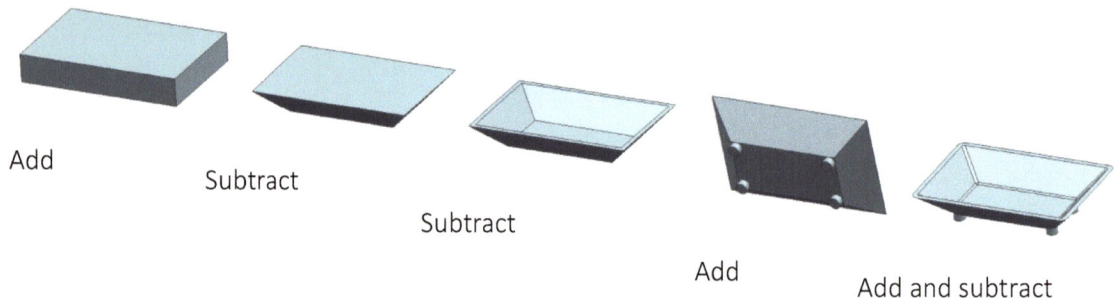

FIGURE 44

On a higher level, solids can be created using sketch entities. A sketch entity is usually a collection of lines, arcs, splines, points, dimensions, and constraints that are all strung together to create a profile. In many cases, it is a closed shape, see Figure 44. Once you have created a sketch, the sketch can be extruded along some axis, revolved about an axis, or driven along some other curve or set of curves.

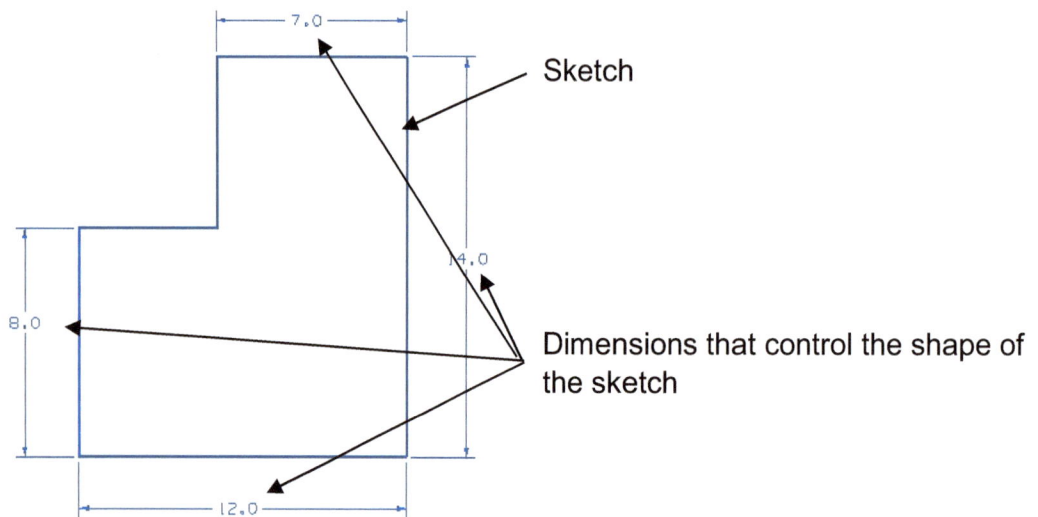

FIGURE 45

In the case of an extrusion this usually forms a solid with capped ends. In the case of a revolve this usually results in a solid that emulates a turning operation. Both ideas are shown in Figure 45.

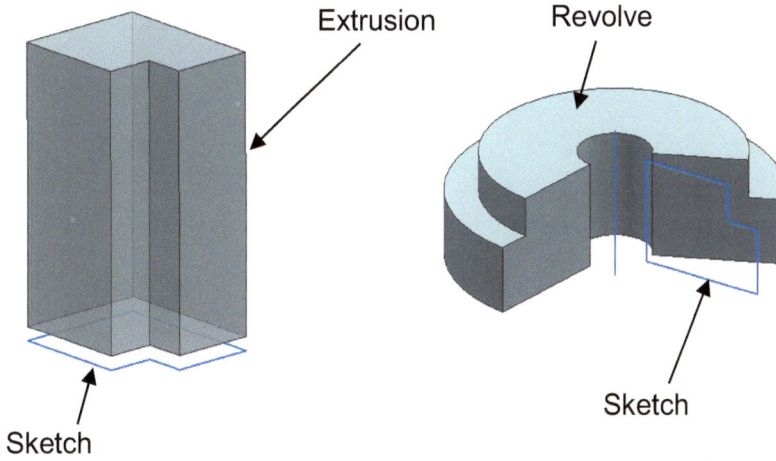

FIGURE 46

Lastly, in the case of a sketch driven along some path this usually results in a solid that is the equivalent of a piece of material that has been drawn, then formed like in Figure 46.

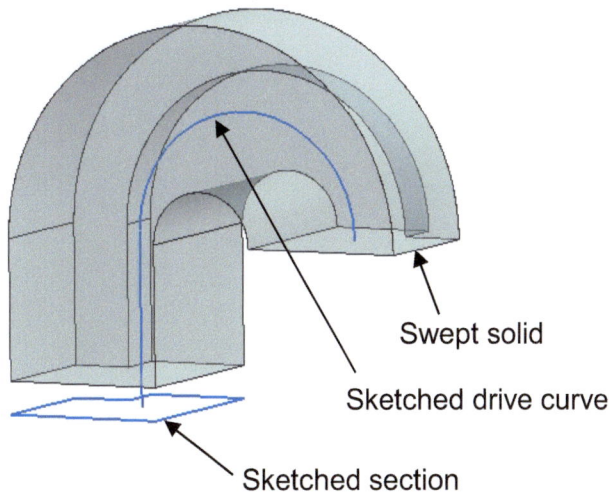

FIGURE 47

It can be easy to understand solid modeling if you think about how operations would be achieved in a manufacturing shop. For example, if you had a flat, square piece of wood and you wanted to smooth out the sharp edges you could use a router or a piece of sand paper. In many CAD systems this is called an edge blend or fillet as shown in Figure 47.

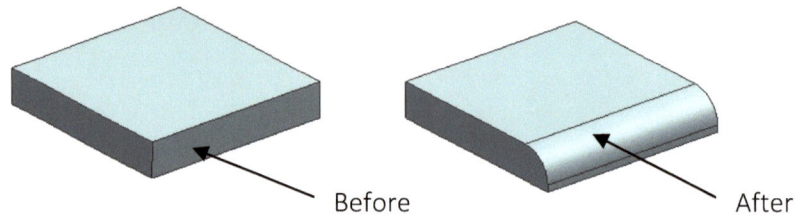

Before   After

FIGURE 48

In most cases the same entity covers exterior blending as well as interior as shown in Figure 48.

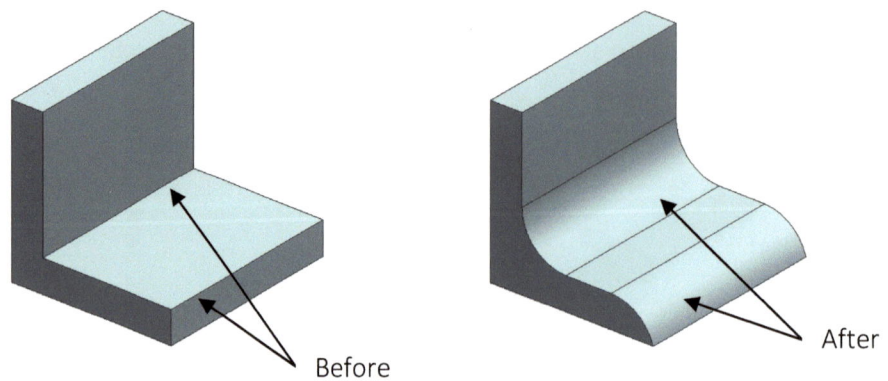

Before   After

FIGURE 49

A similar way to envision the fillet is by rolling a ball against two surfaces and creating a surface of circular cross section everywhere the ball touches.

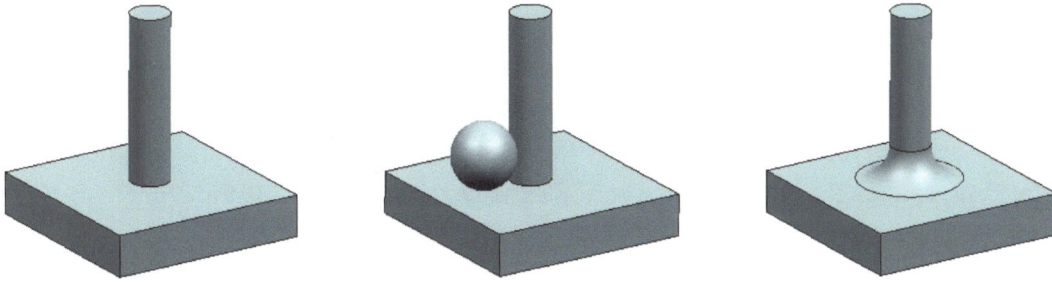

A cylindrical solid is attached to a block.   A fillet is created using a rolling ball.

FIGURE 50

A similar entity is the chamfer, which creates a flat face instead of a rounded one.

Inside chamfer

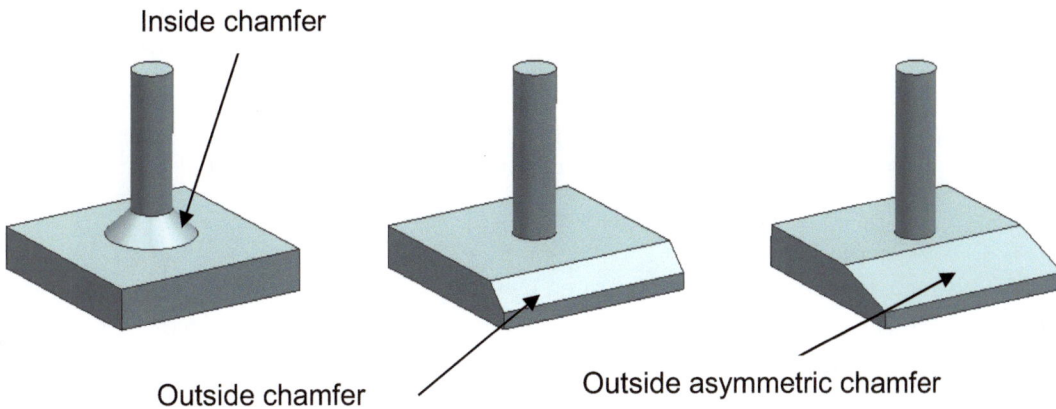

Outside chamfer   Outside asymmetric chamfer

FIGURE 51

Many Solid Modeling CAD programs follow a parametric, features based paradigm. That is, models are created by adding many features together in a sequential format that are driven by parameters. For example, you can create a model of a beer mug starting with a cylinder that is 8 inches high and 4 inches in diameter as shown in Figure 51. The cylinder is the feature and the dimensions are the parameters. Next, one performs a "hollow" or "shell" operation so that the cylinder has room for the beer. The shell command specifies the wall thickness at 0.5 inches. Then a handle is added using a sketch and a sweep command. The list of operations is then, cylinder, hollow, sketch, sweep, and add. This list of operations is called the Constructive Solid Geometry tree (CSG tree). It is the backbone of feature based parametric modeling.

Along with the CSG tree there is a database of every numerical value that was entered as the geometry was created. The beauty and the power of the feature based parametric paradigm is that any time a design change is required, all you have to do is change one of the underlying parameters in the data base and the entire model is recreated. In the case of the beer mug, when one wants a larger mug the parameter that captures the original 8-inch height is changed from 8 to 12. All the other features then update. Figure 51 illustrates every step of the example above.

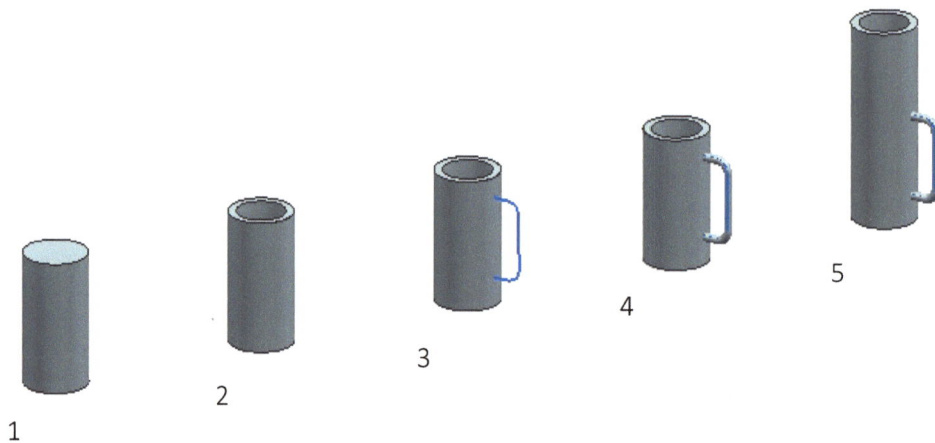

FIGURE 52

In many CAD systems, modeling is greatly facilitated by using constrained sketches. They provide the user with the ability to truly capture design intent from a 2D perspective at first, then 3D later. The entity is created on a default flat plane, or datum plane, and used in subsequent operations to create 3D geometry. A sketch captures the design intent by allowing a user to dimension and/or constrain the various entities that make up the sketch. Each dimension of each sketch shows up in the database as a parameter that controls the 2D shape in the same way parameters control solid features. When you need to change the shape of the sketch you may access the dimensions and the entire shape updates.

The true power of a sketch is its ability to capture more than just dimensions. For example, if you are creating a model of a window, and you want its height to be 1.618 times the width (the golden rectangle relationship), you can easily enter that mathematical relationship into the database. The beauty of this ability is that it allows you to iterate on the width of the window and have the height automatically update for any width. The automatic nature of constraints and dimensions is what provides the greatest ease when creating a sketch and subsequent model. To illustrate the point further, consider the two simple shapes in Figure 52. The one on the left is dimensioned with the inner profile dimensioned from the left-hand side. Due to these numerical values, the overall width being 60, the width of the window being 30, and the distance from the left-hand side being 15, the widow is centered. The contour on the right has a centering rule constraint for the spacing of the window instead. When a design change is made in which the overall width is made larger, the shape on the left doesn't update properly as shown in Figure 53. Therefore the sketch on the right captures the true design intent because the shapes are always going to be centered. The sketch on the left does not.

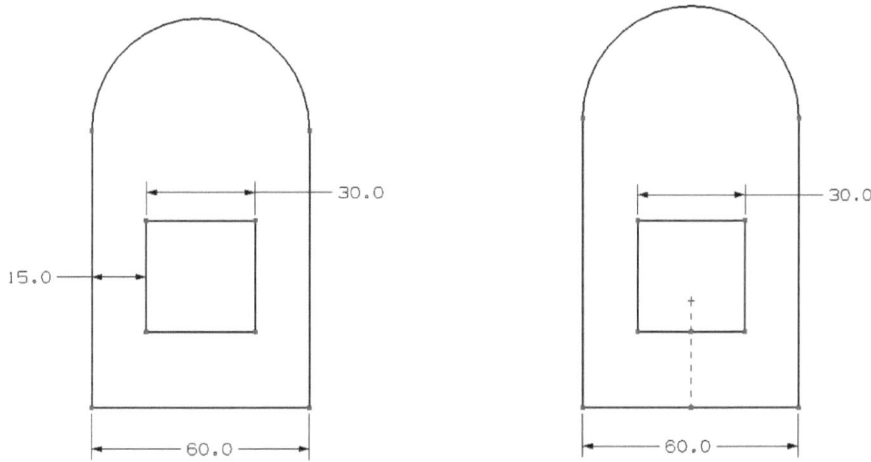

Two identical shapes dimensioned differently

FIGURE 53

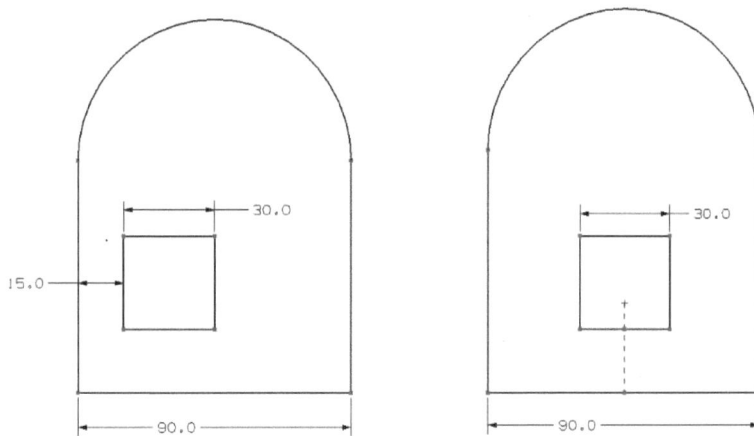

Two contours that react differently to a change in overall width

FIGURE 54

In addition to sketches that capture design intent, modern solid modeling is augmented by smart features such as holes, bosses, pads, pockets, ribs, and more. What makes a smart feature smart is its ability to capture not just the geometry that was intended when it was first created, but the true design intent of its creation as well. For example, the hole in the solid shown in Figure 54 is intended to have a depth of half the thickness of the block it is drilled into. When the block thickness is changed, the hole depth automatically updates to half the new thickness.

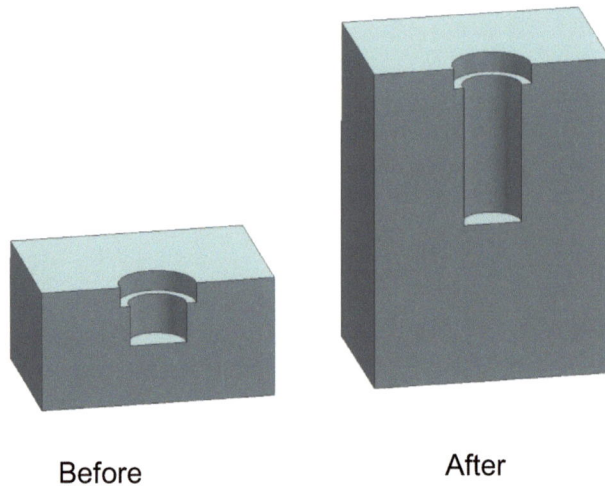

Before　　　　　　After

FIGURE 55

Other smart features are the boss, pad, keyhole, threaded hole, and groove, to name a few.

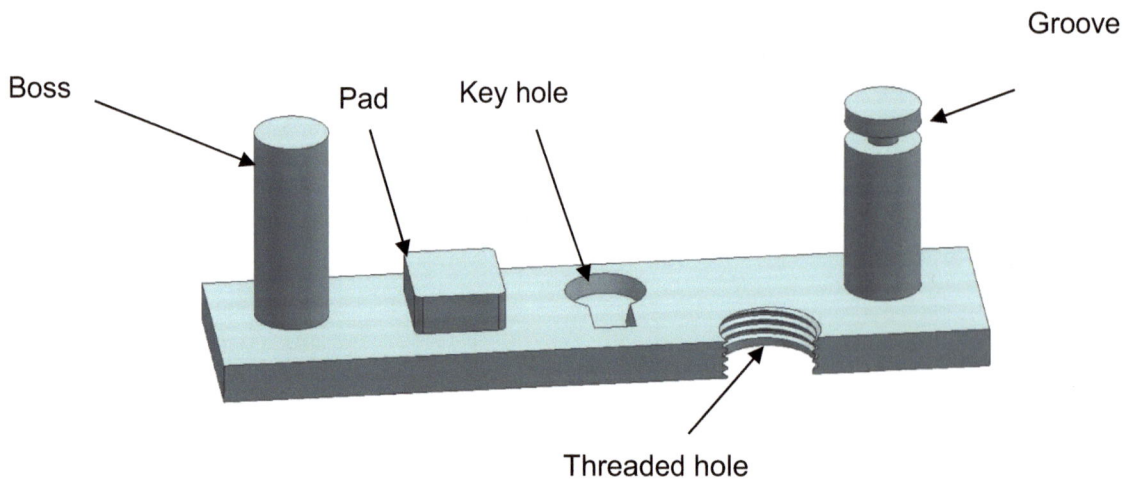

Boss　　Pad　Key hole　Groove

Threaded hole

FIGURE 56

REFERENCE ENTITIES

The creation of complex solid models with challenging geometry often requires the use of reference entities such as datum coordinate systems, planes, axes, and datum points. These help to position other entities such as sketches, curves, etc. Modern CAD systems allow you to create these entities in almost every conceivable way. For example, a datum plane can be created in an offset from a face, through three points, through two non-parallel axes, one point and an axis, tangent to a cylindrical face, and more. Parametric datum planes update automatically and serve as the back bone for many complex models. For example, a propeller model is made by creating a series of datum planes thru a cylinder, then using the last datum plane in the series to create a propeller blade. The pitch of the blade is controlled by the angle of the datum plane. Each entity is necessary for the creation of the next. See Figure 56.

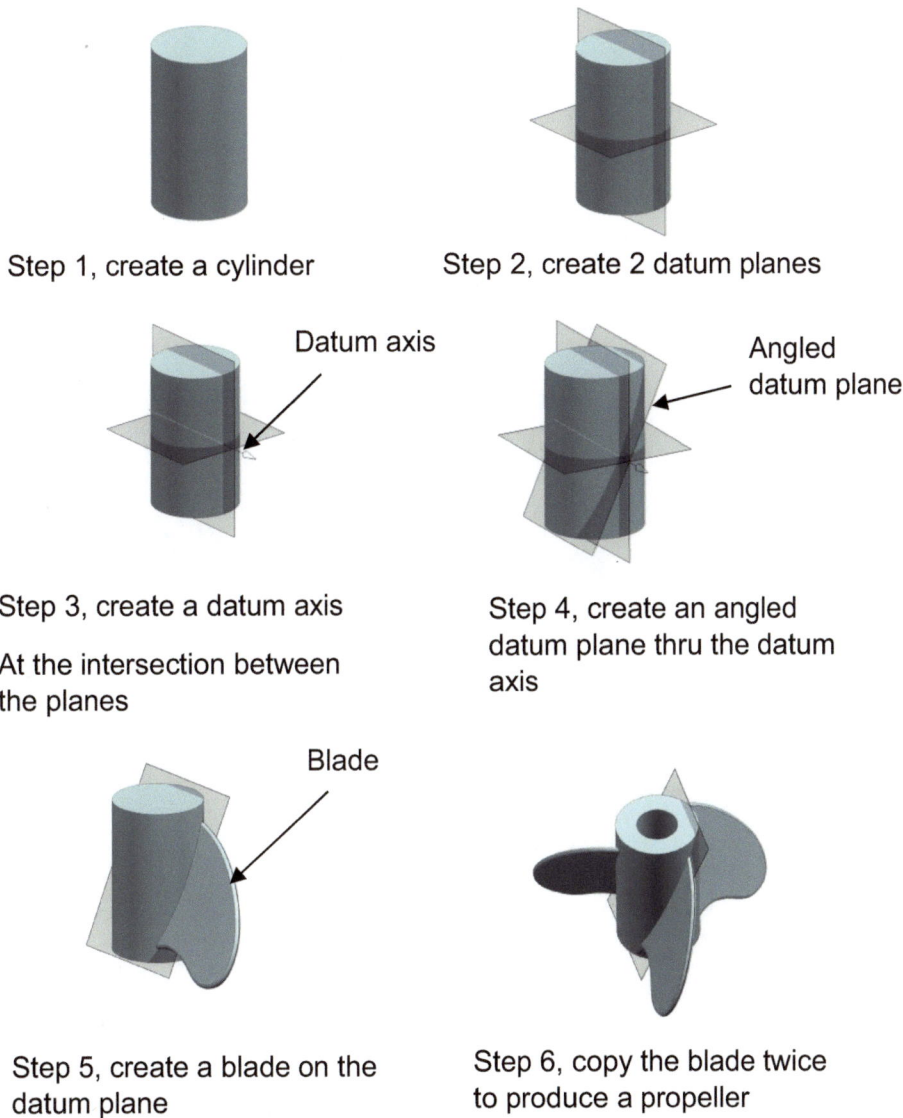

Step 1, create a cylinder

Step 2, create 2 datum planes

Datum axis

Step 3, create a datum axis

At the intersection between the planes

Angled datum plane

Step 4, create an angled datum plane thru the datum axis

Blade

Step 5, create a blade on the datum plane

Step 6, copy the blade twice to produce a propeller

FIGURE 57

Datum curves are created in 3D space to serve as the skeleton for geometry. The example below shows the creation of a curve mesh surface that is created from a series of curves.

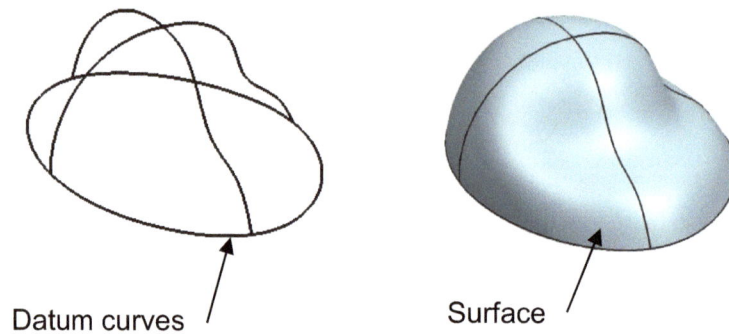

Datum curves   Surface

FIGURE 58

Datum points can be placed anywhere in 3D space in almost any conceivable way. The points generally support the creation of other 3D geometry. For example, points can be distributed on a surface to serve as the locations to make blister features.

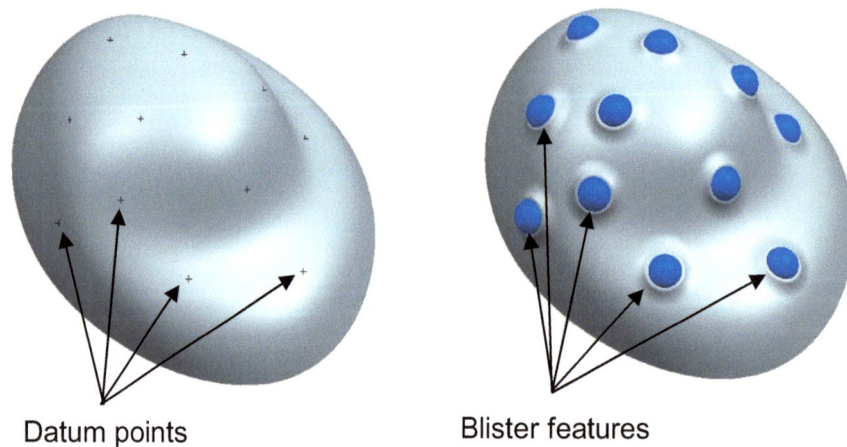

Datum points   Blister features

FIGURE 59

Along with datum axes, planes, points, and curves most good solid modeling systems give you the ability to create datum coordinate systems. A datum coordinate system is essential when positioning solid models in 3D space. The datum coordinate system is an entity that allows you to position it by keying in offsets from the global coordinates system. When geometry is created on datum coordinate systems the geometry is moved when the parameters of the datum coordinate system are varied.

EXPRESSIONS AND VARIABLES

Most modern solid modeling systems allow for the creation of expressions or variables. You can assign a numerical value to a variable name such as thk =0.25. As you continue creating various features you use the expression and the geometry that you create, which is then associated to that expression. For example, a simple enclosure is created with a wall thickness of thk =0.25. When the ribs are created to strengthen the enclosure the rib thickness is related to the variable (0.65 × thk). The design intent that is captured is the desire for the rib to always be thinner than the overall thickness of the enclosure thereby avoiding the sink problems that can occur during an injection molding process. This ability is useful when a design change is needed. When it is determined that the overall wall thickness needs to be greater, and thk is re-set to 0.35 all the ribs update automatically. In this way, expressions serve the purpose of capturing the design intent and the numerical relationships between various features.

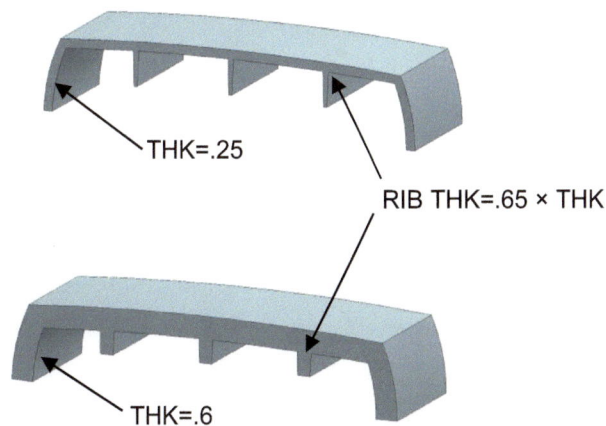

THK=.25

RIB THK=.65 × THK

THK=.6

FIGURE 60

Expressions are also great for driving geometry that follows a mathematical function. For example, the following set of parametric equations creates the crazy double helical shape below.

amp1=.5

amp_n=50

n=4

p6=.25

p7=0

pitch=3

r=5+amp1*cos(amp_n*t*360)

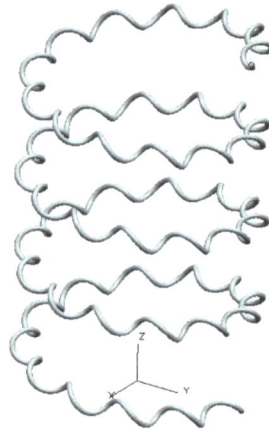

FIGURE 61

SURFACING

Surfacing is extremely useful whenever creating shapes like airfoils, human forms, consumer products, sporting goods, toys, or other things in a model. Perhaps the most common way to create a surface is by making a number of sections that vary in shape and "stretching" a skin over them like an airplane wing. See example in Figure 62.

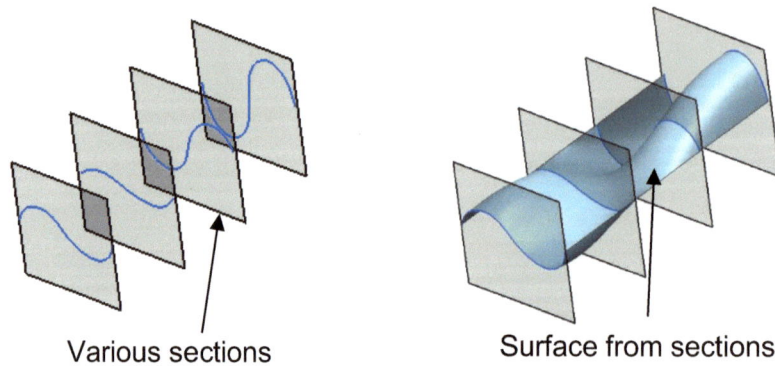

Various sections   Surface from sections

FIGURE 62

When a surface has to be controlled to a greater degree than planer sections can afford you may need to supply curves in between the sections. When a surface is created with a set of section curves and a set of curves that connect the section curves it is usually called a mesh surface.

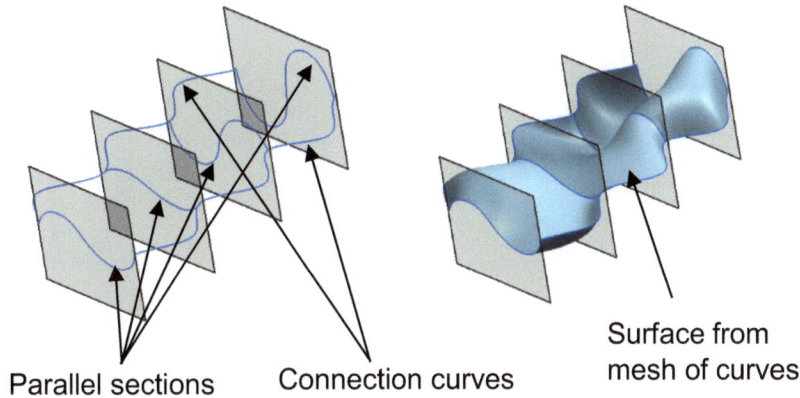

Parallel sections   Connection curves   Surface from mesh of curves

FIGURE 63

Many forms in engineering are constructed from sections swept along a path. Whether you are creating a diverging fuel nozzle, or a sound attenuating device, you need a way of creating surfaces along some central path or drive curve. In the example below, two semi-circular cross sections of different radii are swept along a spiral. The result is the surface shown.

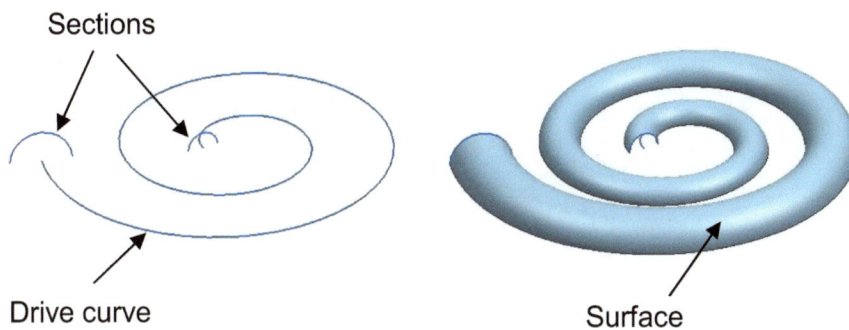

Sections

Drive curve   Surface

FIGURE 64

In most solid modeling systems there is a surface that can be used to make a transition between two existing surfaces of different shapes. This can be used, for example, to make a model of the duct work behind the dashboard of your car or perhaps a model of the latest hip implant. The procedure involves selecting the edges of two separate surfaces and instructing the transition surface operator to create a smooth and continuous surface that flows from one edge of one surface to the other while maintaining tangency at the take-off edges. See example below:

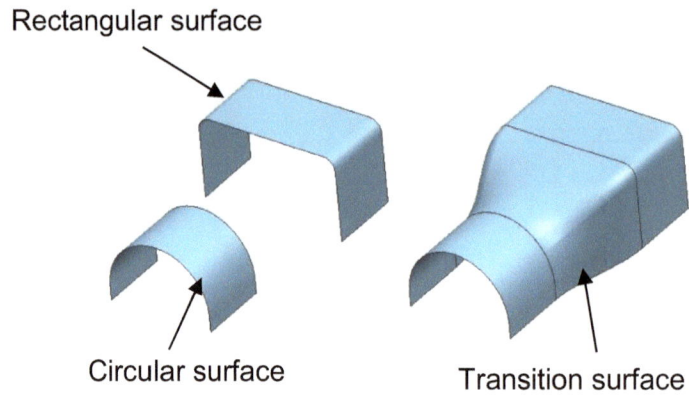

FIGURE 65

In many cases, surfaces are created from real objects in the physical world. It's usually a multistep process where an object is scanned with a laser or a coordinate measuring system to produce a point cloud. The point cloud is most useful when it follows certain rules such as the points being arranged in horizontal rows with equal numbers of points in each. See Figure 66.

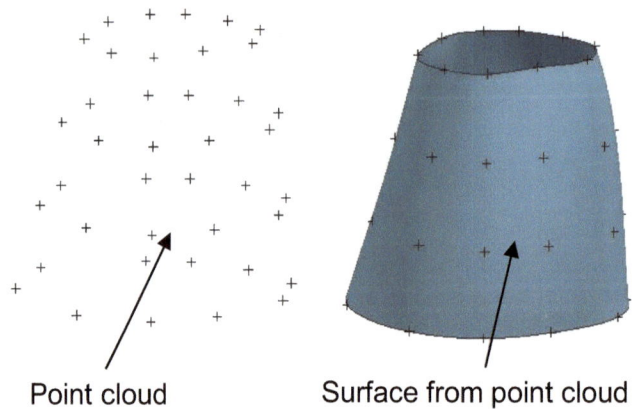

FIGURE 66

A surface differs from a solid model in that it has no mass, no inside or outside, and is infinitely thin. Surface modeling is a useful tool when modeling because there are many ways to convert a surface or set of surfaces into a solid. For example, you can add thickness to a surface, automatically creating a solid.

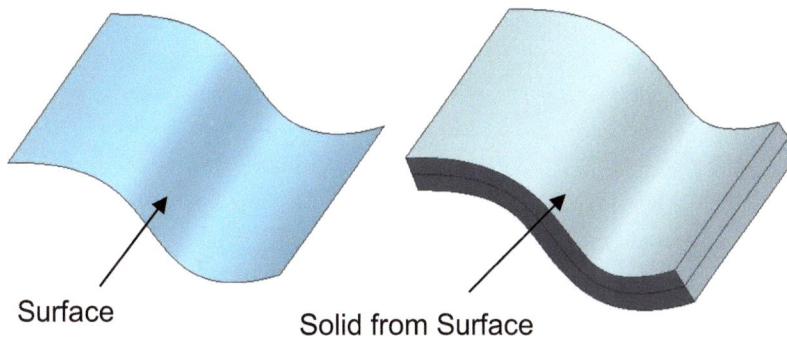

Surface   Solid from Surface

FIGURE 67

You can also take a number of surfaces and stitch them together on the edges. Some programs refer to this operation as a quilt or a sew. The collection of surfaces has to be air tight (all edges must be aligned) to become a solid. See Figure 68.

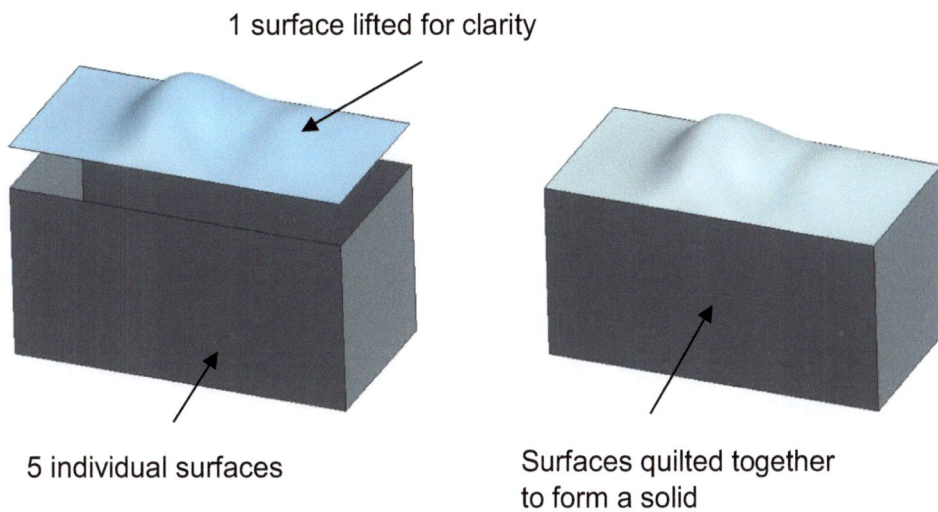

1 surface lifted for clarity

5 individual surfaces

Surfaces quilted together
to form a solid

FIGURE 68

A surface can be used to carve away at an already established solid body. In an operation that can be described as a cut, a trim, or a carve, a surface is placed in a strategic location relative to a solid body. The operation is performed, and one side of the solid is cut away. This is a very useful technique for certain consumer products that have a number of highly sculpted regions such as speaker locations on a radio, the front face of a fax machine, or the input card orifices on a laptop computer.

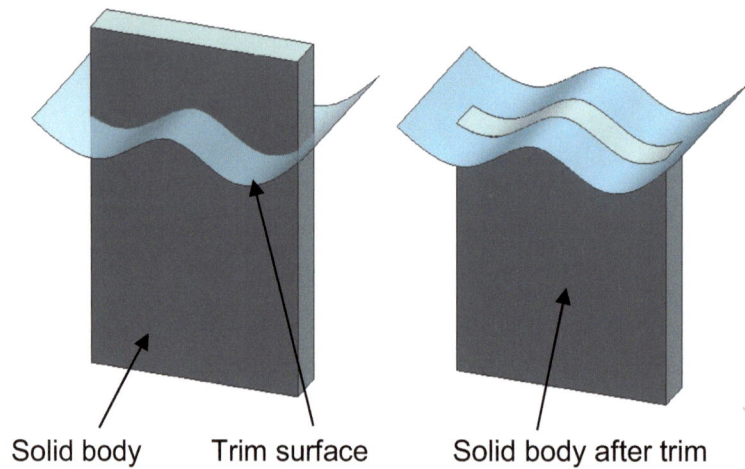

Solid body   Trim surface   Solid body after trim

FIGURE 69

SHEET METAL

Modern high-end CAD systems afford a number of sheet metal functions. The user normally starts with a flat panel with a uniform thickness. Various sheet metal features are added on until the component is done. The main advantage to using sheet metal features is that they carry the extra information required to produce an accurate flat pattern when the design is done. The example below shows a piece of sheet metal that has been created very simply and then flattened.

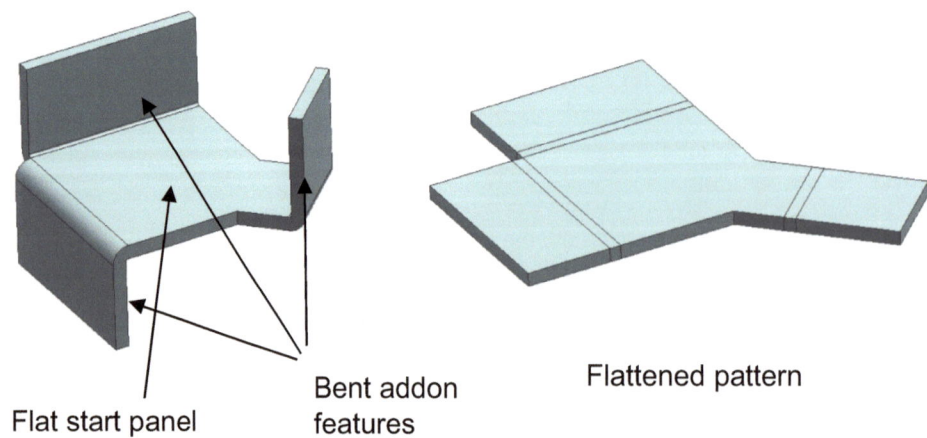

Flat start panel   Bent addon features   Flattened pattern

FIGURE 70

The sheet metal options also include a number of other common sheet metal contours such as louvers, dimples, bends, and cut-outs, see below:

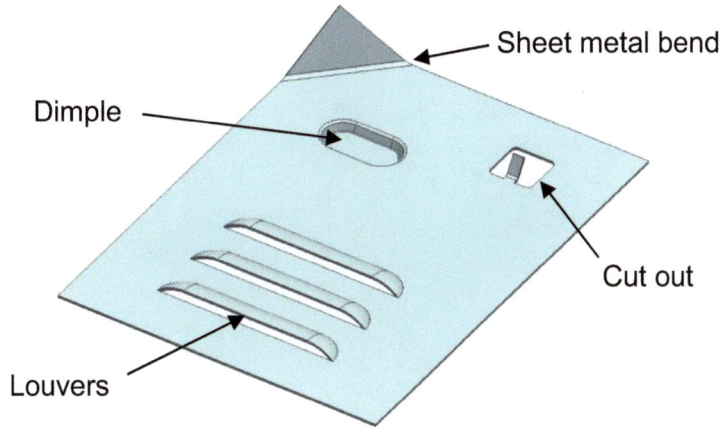

Sheet metal bend

Dimple

Cut out

Louvers

FIGURE 71

ASSEMBLIES

Once you have used the techniques mentioned above to create individual components, you may then begin to assemble them into an assembly. Typically a simple assembly is a number of separate part files with separate models that are being "collected" by one assembly part file. For example, the assembly below of a simple drawer unit shows each and every separate component in its own part file.

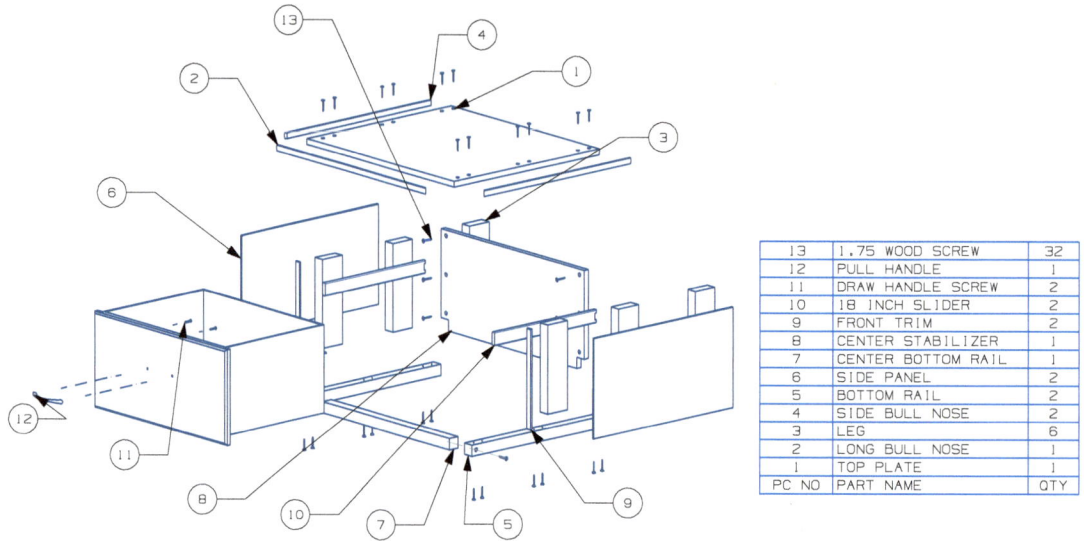

DRAWER ASSEMBLY

FIGURE 72

13	1.75 WOOD SCREW	32
12	PULL HANDLE	1
11	DRAW HANDLE SCREW	2
10	18 INCH SLIDER	2
9	FRONT TRIM	2
8	CENTER STABILIZER	1
7	CENTER BOTTOM RAIL	1
6	SIDE PANEL	2
5	BOTTOM RAIL	2
4	SIDE BULL NOSE	2
3	LEG	6
2	LONG BULL NOSE	1
1	TOP PLATE	1
PC NO	PART NAME	QTY

The main file in this case that calls in all the other part files and automatically opens them, is called the Drawer Assembly which is shown below:

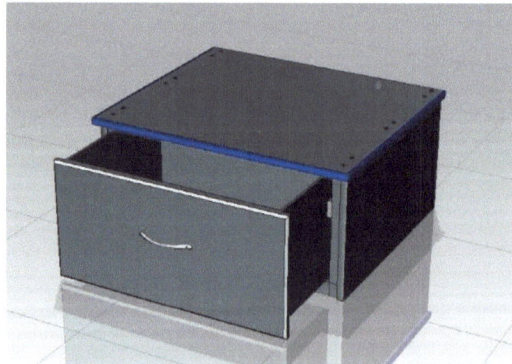

FIGURE 73

The assembly file contains all the information that associates each component to each other. There are two main ways an assembly defines the geometric relationships between components: 3D location and geometric constraints. When components are related by 3D location they are positioned relative to a central or global coordinate system. It is a lightweight method, but it doesn't automatically adjust the locations of the components when one in the group changes shape. For example, a three-stage rocket has a first stage that is 10 feet long and a center stage that is 5 feet long and a final stage that is also 10 feet long. If each component is stacked in the right location, one on top of the other, the entire rocket is 25 feet high. With the 3D location method, should the center stage grow to 10 feet, the final stage doesn't necessarily relocate to the correct height of 30 feet. The stages don't "feel" each other. There will be an overlap either between stage 1 and 2 or 2 and 3 or both. In this example it is very easy to see, but in more complex assemblies it may not be.

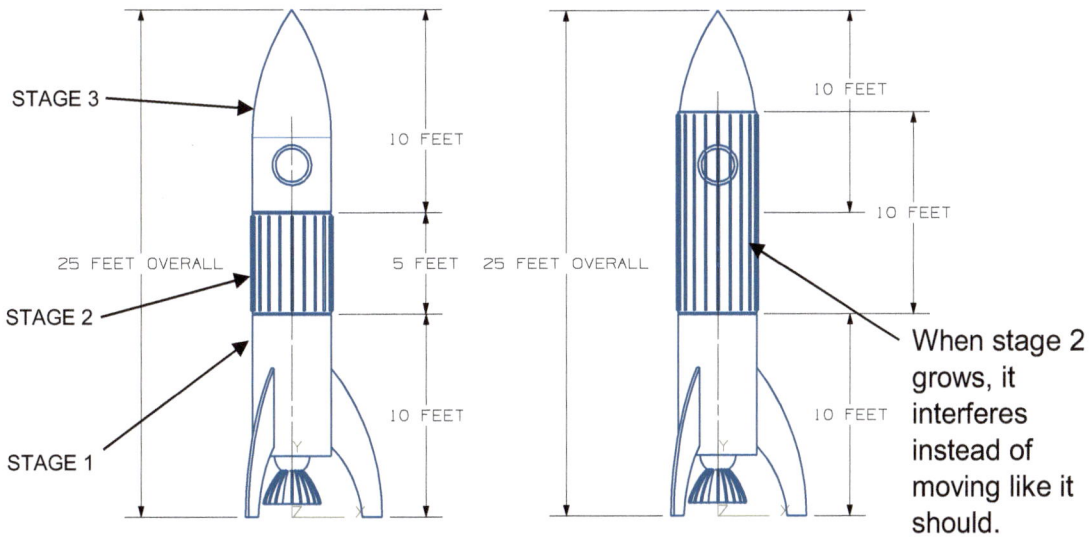

STAGE 3

10 FEET

25 FEET OVERALL

STAGE 2

STAGE 1

5 FEET   25 FEET OVERALL

10 FEET

10 FEET

10 FEET

10 FEET

When stage 2 grows, it interferes instead of moving like it should.

FIGURE 74

When assemblies are put together using geometric constraints, surfaces, edges, centerlines, planes, and other 3D entities are related to each other. For example, the top of stage 1 of our example assembly would have a relationship with the bottom of stage 2 and the top of stage 2 would touch the bottom of stage 3. See Figure 75.

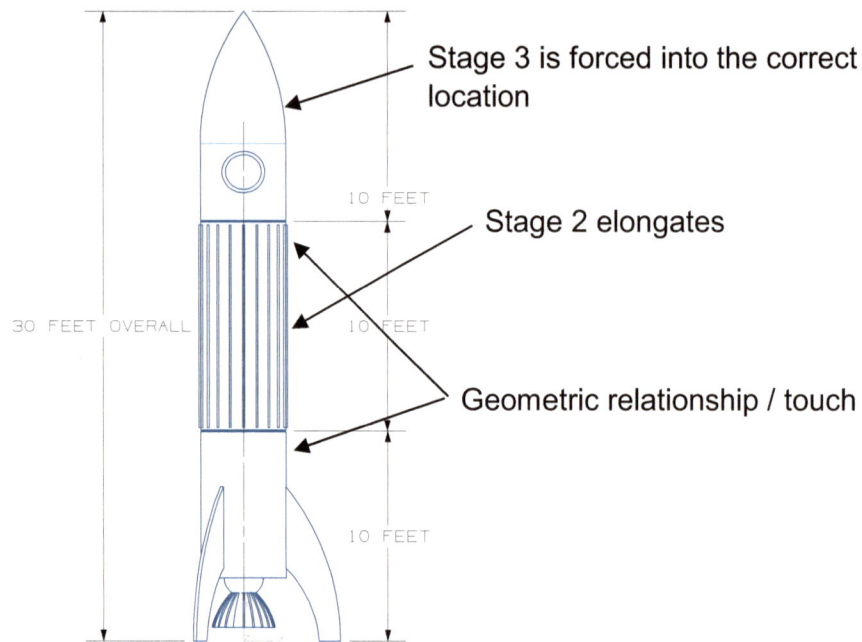

Stage 3 is forced into the correct location

Stage 2 elongates

Geometric relationship / touch

10 FEET

30 FEET OVERALL   10 FEET

10 FEET

FIGURE 75

At first glance it would seem that using the "constraints" method is the way to go, but there are many other considerations. Using the constraints can get very difficult when you are working on an assembly with many components. When many different groups of people are working with a large assembly, it adds on another level of complexity that makes the global coordinate or 3D location method much easier. Each contributor knows where their component has to end up in 3D space, which enables a lot of work to be done without the overhead of having all the other components loaded.

Weather assemblies are performed with the global coordinate method or the constraints method; users still get the benefit of collaborative work and part re-use. It is very common for designers who are working on systems such as consumer products or machinery to save time by downloading models of purchased components right from the internet. Web sites like www.McMaster.com and www.Digikey.com make available models for many of the components they provide as a free download. The models come in a variety of formats and all major CAD modeling programs are compatible with one or more of them.

The assembly's paradigm makes it possible for rapid design creation since you can use entire sub-assemblies from previous projects. When the situation allows, a good designer can use various previously incorporated components and benefit greatly from the fact that the drawing is probably already there along with the specification work as well as test data, etc.

ADVANCED MODELING

There are many other modeling techniques that don't fall into any particular category, but must be present for any high-end CAD system to be competitive. To name a few of these: driving geometry with a spread sheet or program, linking the parameters of one part file to another, borrowing features and or groups of features from one part file for another, and performing advanced logical functions on geometry (such as when the length of something gets to a certain limit, new features turn on) are all useful to have. A lot of the advanced functionality in solid modeling is the ability to manipulate difficult surfaces. A non-inclusive list is as follows: matching edges of various surfaces, extending surfaces, changing the shape of surfaces by manipulating the underlying poles, trimming and un-trimming surfaces, performing non-uniform scale on surfaces, and more. A powerful solid modeling program gives you an ability to create libraries of your own features. For example, an aerospace company designs a product with a certain standard flange with two standard holes through it. Hypothetically we can call it a "rupert flange." A user builds a parametric rupert flange and puts it into the library for all those in the company to use with slight variation. Now anyone who needs a rupert flange can access the library, input their dimensional requirements, and then place their special copy of the rupert flange directly on their geometry. See Figure 35.

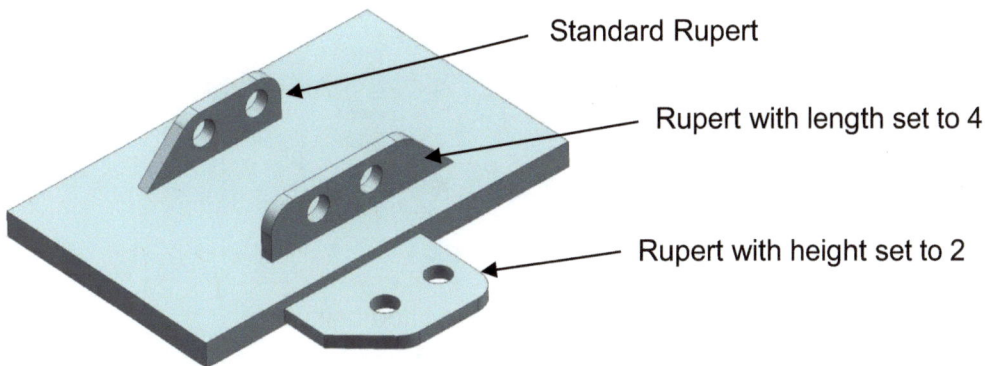

Standard Rupert

Rupert with length set to 4

Rupert with height set to 2

FIGURE 76

EXPLICIT-PARAMETRIC MODELING

When solid modeling was new, everything was modeled explicitly, meaning that the dimensions used to create geometry were no longer accessible once the geometry was created. For example, if you made a 3 by 3 by 3-inch cube, once the creation operation was over, the only way you could change the size of the cube was to perform some type of transform on it. You couldn't access the original data and make a change and have that change cascade trough that solid and subsequent solid features. Explicit solid modeling was the state of the art for years until sometime in the late '80s and early '90s when the parametric paradigm swept over and dominated the world CAD establishment. Once the parametric paradigm caught on, many of the most modern programs abandoned explicit modeling all together.

One of the latest and most exciting developments in solid modeling is explicit-parametric modeling. Explicit-parametric modeling is an incredibly powerful combination of the two techniques. With explicit-parametric modeling a user is allowed to use a powerful set of parametric entities that essentially overwrite the previous parametric entities. In essence, the user is allowed to change the design intent of a particular solid model without in-depth and detailed consideration of what had been done previously.

For example, a model of a speaker housing on an alarm clock has a flat horizontal face on the top where the user is meant to pound their fist on it in the morning. In a redesign to make the experience better, the designer decides to have a little fun with the top surface and make it tilted a bit toward the user. If the software that the designer is using doesn't have explicit–parametric capability the job can get complex and lengthy. They'll have to go back in the database and find the original revolve that was responsible for the top, then somehow reorient it and hope that all the little hole features and the subsequent chamfers update. However, using the explicit-parametric operation, the designer just highlights all the surfaces to be changed then instructs them to tilt about the axis. The change is explicit in that it partially overwrites the design intent and features that were created before. If the hole pattern or any of the other features that preceded the tilt are changed it will cascade through and update. The change is also parametric in that if you go back in the data base and change the numerical value of the tilt angle, all the surfaces will re-tilt to the new angle.

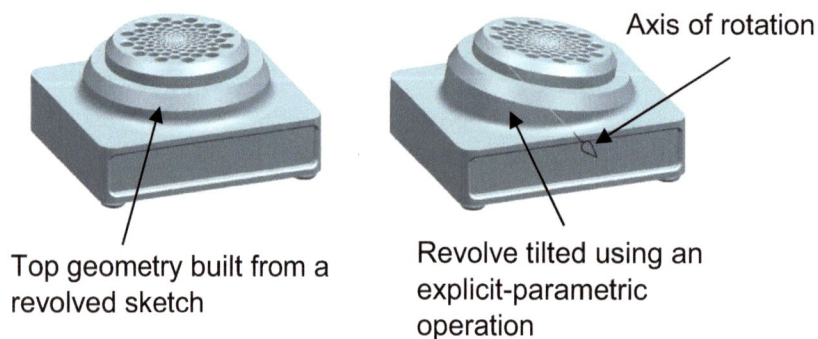

Top geometry built from a revolved sketch

Axis of rotation

Revolve tilted using an explicit-parametric operation

FIGURE 77

MODEL VERIFICATION

All of the great solid modeling systems offer some type of model verification. As a user is creating product definition data, there is a cadre of data verifications that are required. These range from measuring the distance between two points to finding out the full mass properties of a complex model. For those who are working with geometry that will be injection molded, there is a way to check all the surfaces to ensure that they have at least the minimum draft on them. There are ways to find out the optimum parting line of a model that is to be cast, and ways of finding out how a surface will reflect light when it's a real product. There are even ways of finding out how and if components in an assembly are interfering. For example, if you had a peg that was 2.5 inches in diameter inserted into a hole that is too small, let's say 2 inches in diameter, the model will highlight the surfaces that are interfering and even create a separate interference solid upon your request. See Figure 37.

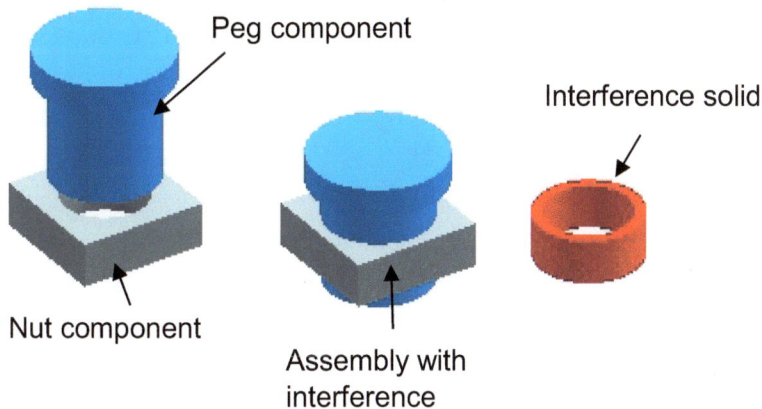

FIGURE 78

ASSOCIATIVITY AND CONCURRENT ENGINEERING

When a CAD model is created, it typically captures the 3D geometry of a component otherwise known as "the shape data." It usually doesn't capture the non-shape data such as surface finishes, geometric tolerance, heat treat, or other special manufacturing instructions. This "non-shape" data is usually captured on a 2D drawing. The 2D drawing is linked "associated" to the 3D part. Every powerful CAD system affords the user the ability to create a separate file for the 2D drawing that in essence has the 3D geometry part file as a one-level associated component. Designers can place 2D views and their dimensions into drawings. The views are still associated with their 3D parts and can be updated when parts are edited. This ability allows two users to work on a job at virtually the same time. The 3D designer begins the job and shortly after, the 2D drafts person begins the drawing. As the 3D designer improves the design the 2D drafting model receives updates automatically.

The associative nature of modern CAD systems affords concurrent engineering and collaboration to those responsible for all the other product design functions such as stress analysis, manufacturing, testing, kinematic analysis, mold flow, and anything else. Most competitive CAD packages are sold with options to do all of the above in one seamless integrated package. These systems, as of the time of this writing, can be purchased for prices from $1200 to ten times that depending on the options you are interested in.

PRODUCT LIFECYCLE MANAGEMENT

Making progress with other engineers, designers, manufacturers, marketers, and the like can be a very difficult and challenging process. There is a "holy grail" of product design that most of us in the business are trying to achieve. Products are defined in such a way that they optimize the input of all functional groups. In essence, it's the widget that looks exactly like what marketing would want, as functional as engineering would like, as easy to manufacture as manufacturing would like, and even set up to be recycled as easily as any product can be as environmentalists would like, and it sells like hotcakes and makes us all wealthy. In order to achieve this "holy grail," each product has to have much consideration from each of the contributing groups as early and as often as possible.

The essence of collaborative engineering is managing the input from all these people with different focuses and ideas. For this, engineers and product designers use Product Lifecycle Management (PLM) systems. A good PLM system has the incredible ability to allow access to Product Definition Data (PDD) no matter where in the world team members are. The premise is "if all the members of the cross functional groups have access to all the data all the time, the different phases of product development will be in parallel instead of in series." For example, if the manufacturing engineer, who usually doesn't see the design until the drafting department throws it "over the wall," can gain access to the data early, and helps define a better shape for a certain casting that is to be produced, this will mean greater gains in manufacturability. The "holy grail" is in sight as long as we all communicate as up front and as effectively as possible.

Product Lifecycle Management (PLM) is everything that is necessary to streamline the innovation and management of product definition data (PDD), old and new. PDD can be an array of CAD part files, specifications in the form of pdfs, Word documents, manufacturing and inspection instructions, user manuals, etc. As this data is created using CAD systems, it may be augmented by other data found in libraries, separate folders, and other places. It all must be organized so that the right data is given to and/or created by the right person at the right time. Whether it's a simple toy or a complex jet fighter, it is critical that the PDD be organized, never misplaced, and never accidentally substituted or altered by the wrong procedure or person. As a product is defined in its life cycle, a good PLM is able to not only organize the data, but also keep track of all the procedures that were put in place to create the data. A good PLM system can tell you very quickly the current definition of a product and its history from cradle to grave. It gives you the power to do it all and in doing so, provides a huge advantage to those who are innovating in any competitive market. Most CAD modelers have a PLM system that can be purchased with the CAD software and most PLM systems can manage the files that come from a CAD package that is not in the same product family, but as of the time of the writing this there is usually something lost in translation.

THE FUTURE OF CAD

Talking about the future is always risky. The quantum leaps that occur over time, especially in any business that has to do with software and computers, make predictions extremely difficult. However, what is apparent is what CAD users want. Hopefully what CAD users want is what the industry will provide sooner or later. CAD users all over the globe have difficulty finding the commands they're looking for. It is extremely important for solid modeling systems to get better at organizing and streamlining the user interface so commands are easier and easier to find. To some degree, the most powerful systems are at a disadvantage to the lesser ones due to the sheer volume of commands and abilities that they have over lesser systems. It can be confusing, especially for new users.

Another general thrust in CAD systems comes from the fact that most design engineers would love to find some way to dispense with the entire drafting process. Indeed, high-end systems already have the ability to capture things like geometric dimensioning and tolerancing, surface finish callouts, and non-shape entities right in the 3D model. In the future this will undoubtedly be made easier and more common.

Another controversial improvement in CAD is the total abandonment of parametrics. There are a large number of engineers who never liked the move to parametic modeling. Commensurately there are a small yet growing number of CAD software designers that are determined to make a non-parametric paradigm that is as productive and easy to perform design iterations with as parametric modeling. The future may yield a solid modeling package that gives anyone at any time the choice to be parametric, non-parametric, or lightweight.

By and large all CAD modelers look at the features and functions of others and make sure that somehow they get everything that everyone else has. In recent years, CAD software companies have bought each other out and made new programs that incorporate everything that the other program had plus what they had originally. It's also a great way to increase the installed base of users. As we go forward, I think that CAD systems of the future will be fewer, far more powerful, less expensive, more widely known, and easier to use as an expert or a novice user. I also think that the more powerful the systems become over time the more exciting and fun they will be to use. More things will be easier to do, with more computing power and capabilities, and we as designers will be asking them to do increasingly more difficult geometry. To some degree designers limit the shape of their designs based upon the limitations of the CAD systems that they use. In the future we will all enjoy more freedom for better designs.

PROTECTING YOUR IDEA

Some friends of mine were working in a company that manufactured a surgical robot. The company kept going for about ten years and never made a dime. They kept going back to investors and getting more money, but they just couldn't get things going well enough. At some point the two of them got laid off, so they came to me and wanted my help with a patentable idea that would assist with knee replacement surgeries.

One of them, who had a PhD in manufacturing engineering, described the idea to me. He had a very thick accent that I had trouble understanding. As he described the idea to me, and I tried to make sense of what he was telling me, I did what many mechanical engineers do: I began to draw sketches of what I thought he was talking about. When he saw them, he realized, "Hey, that's even better than what we were thinking!" That slight communication barrier would normally have hindered the project. Instead, this time it helped make it better. So away we went with the idea that evolved from our conversations and later CAD models we built and techniques we discovered.

The idea as a whole involved taking CT scan data of a damaged knee to use with rapid prototyping in order to make a custom jig that would guide a resection tool during total knee arthroplasty (TKA) surgery. We developed some really powerful algorithms to go along with the technique. The overall advantage that we provided was to ensure perfect alignment of the prosthetic knee joint geometry during the surgery.

When we had developed the idea to a certain level we applied for a patent. My design firm supplied all the funding since the other two partners had just been laid off. One of the partners was chosen to go out and sell the idea for what we imagined would be a huge amount. I didn't go sell it myself because I had a company to run. To my surprise, he came back empty handed.

At some point both other partners were supposed to perform other engineering tasks for me as we waited for some response to balance out my financial contribution, but that didn't end up happening. Weeks turned into months and eventually the project fizzled out. The two other partners got jobs in other companies, and the whole project was put on hold. I had a business to run and many other priorities. In time, I lost track of the patent and the other partners.

A few years later, and to my surprise, one of them called me and told me that he was interested in buying me out of my share. I asked him if he had located a buyer, and he said no, he hadn't, but that he just wanted to give me some money because he felt bad about how everything worked out. I was a bit skeptical, but I was glad to take the eight thousand dollars he had offered, which was at least something more than the nothing I would get if I just kept sitting on it. I took the money and smiled as he walked out the door. I forgot about all of it until several years later when I got a call from a lawyer from Stryker Medical Company.

The attorney on the other end asked me if I would be willing to help recertify or perform some type of maintenance on a patent that appeared to have my name on it. I immediately knew he was talking about the TKA machine, so out of curiosity I asked how much Stryker purchased the patent for. The lawyer said the he could not divulge the purchase price but all the details were on a website. When I made my way to the URL I was shocked. The article stated that the patent had been purchased by Stryker Medical for a sum total of $63 million dollars with a future payment of $30 million. I was amazed and delighted that I had the privilege of helping develop an idea that had that much value to folks. I was proud that something that I had a part in was helping people get better and more reliable surgeries. It was a great learning experience and ego booster, too.

Naturally there was a part of me that wondered if there was any way I could be compensated a bit more for the idea that I'd worked on. I didn't pursue it very hard, though. I felt that the decisions that I had made were good ones given the information I had at the time. Also, in the meantime I had done pretty well for myself financially and otherwise. The overall impact of the experience was one of validation. In the end it indicated to me that the creative ideas that I can help come up with are worth many millions, and more importantly can provide better care for many people. I find it motivating, and I want to make sure that the experience isn't a one-shot flash in the pan. Next time, I shouldn't sell the idea for such a small price. The experience taught me to place more value on my ideas and encouraged me to work in the direction of intellectual property, getting more of my own designs patented and developed.

IDEA KILLERS

For the most part, those that come up with great ideas are those who have been thinking creatively all their lives. Out of the relatively few people who are truly creative and have the drive to make something of that creativity, there is a small subset who decide, or are compelled, to protect their ideas to the degree that they can reap the financial reward. Assuming that you are a person who truly comes up with great ideas, or who is just about to, you will need to protect them. In order to protect them, it is important to consider the threats.

The biggest threat to your ideas is you. Most of us have obligations and commitments that are constantly competing for our time. It's safe to say that there are many ideas that we all come up with that are knocked out of our minds as soon as those other obligations come knocking. Our ideas may or may not be great, but regardless of quality they are ignored in lieu of the necessities of a busy life. It's a self-inflicted wound.

Another big threat to your idea is sitting on it too long. You can easily have a wonderful idea that would benefit the entire planet and be too busy to write it down or move it along in any substantial way. Later you will see a commercial or some publication saying that very idea is a full-fledged product now selling for millions. Sometimes the product idea sits until it is no longer needed. Imagine if you had a great idea for a device that would make gas engines more efficient and powerful at the same time, but you waited and waited to do anything about it only to find out that everyone was moving toward electric cars now.

Among the biggest threats to your idea is a lack of confidence. Remember to avoid the trap of telling yourself "if this were a good idea someone would have done it before." It's that line of thinking that kills a huge number of great ideas. This can only be overcome by your belief in yourself and your idea. Be confident!

Another threat to your ideas is when they aren't really yours. When you come up with a great idea and you find out that someone came up with it a few weeks, months, or even years earlier, it's really not your idea, even if you came up with it completely independently. Legally speaking, it's theirs. This is perhaps the biggest threat to your vocation as an inventor. It is not only about who had the idea first, though. As long as the invention is not yet publicly available, patent law bases priority on the first inventor to file a patent rather than the first to invent the idea. Also, relatively few people have the special combination of confidence, creativity, intelligence, resources, free time, motivation, energy, and everything else you need to actually go through the process of getting a new product on the market. So even if you're not the first one to come up with a particular idea, there's a good chance that if you're the first to file for a patent you'll be the one to reap the benefits.

Other people stealing your ideas may be one of the lesser threats compared to your own mindset and timing, but it is still an incredibly important thing to consider when protect your idea. Even just mentioning your idea to someone else can be an issue. There are relatively small numbers of people who are true innovators, and the chance of you having a great idea and accidentally mentioning it to the wrong person who goes off and creates a patent and a product without you is probably small. That said, once other people know about your idea, it's hard to control whether that idea becomes public. If you have an unprotected idea and you go off and broadcast it on social media to the world then you are extremely vulnerable.

DOCUMENT YOUR IDEAS AND DON'T PUBLISH UNTIL PROTECTED

It's important to keep track of the development of your idea. Make sure you have drawings with dates and signatures on them. They could be hand sketches or computer renderings. It's good to have a bound notebook with pages that can't easily be put in or taken out. If you created the idea on a computer program and it's only on your hard drive, print it out with a date and sign it. Put it in a safe or under the proverbial mattress for safekeeping.

When you have a great idea in mind, it may be tempting to share it as soon as possible so the world may know of your genius. However, doing so can kill your idea before it even gets off the ground. Publishing before you are ready brings up a host of issues and opens you up to getting your idea stolen out from under you. The best practice is to guard your idea fiercely and ensure you are protected before you share it with anyone.

"Publishing" an idea may sound like it requires a formal, legal act, but it is actually much less strictly defined than you might think. This can be a dangerous thing if you aren't careful. For example, if you upload a video and share it online, you have "published" the idea. As soon as you publish your idea you can no longer patent it. This is because your published idea can now be used against you as "prior art," which is any publicly available information your invention may be compared against to determine its novelty when you file for a patent. There are a few exceptions to this rule. If the prior art was produced by one of the inventors listed on the patent less than a year before the patent was filed, it may be alright. Even so, publishing before you are ready is a risk you don't want to take.

Along the same vein, don't make a prototype of the invention to show off to clients or anyone else. That's another no-no for later patent submission. Don't talk about the idea with friends, and don't let anyone help you without signing an NDA (a non-disclosure agreement) to ensure that they don't end up making your invention public before you are ready. Even showing your idea to an engineering firm or the like without full protection is risky.

DON'T RUSH DOWN TO THE PATENT OFFICE

Sometimes a patent is not enough. You can easily spend a bunch of money on a patent, submit your idea to a firm that is supposed to help you, and have them steal the design from under you. Even if your patent gets filed, the protections you get from a patent can look laughably meager to groups with more resources and opportunity than you. A patent only gives you the right to sue, but you can't sue without considerable resources, so when you submit your idea to a large company that has a bevy of lawyers, there is still a chance that they will steal your idea. Be familiar with the reputation of the company that you're submitting to. Find out how they treat or mistreat inventors. Also realize that when you patent an idea, it's made public, enabling competitors to read the text of the idea and figure out things about it that may be best kept a secret. There are folks whose job it is to figure out how to "one up" your design, and making your invention public through a patent may give them the means to do just that.

COSTS AND BENEFITS

Getting a patent is costly but worth it, even if big companies can often times get around it. The patent will stop reputable companies and fair-minded individuals from stealing your idea. The knuckleheads may come after you anyway, but it will make it harder for them. It's a matter of money and numbers.

To have an attorney file a patent with everything necessary for it to be worth anything costs a significant amount of money. It costs even more to maintain that patent, and to fight those who would infringe upon it. At the time of this writing, we looked around to see what one might cost. To file a **non-provisional patent**[1], not including other expenses that might come up along the way such as filing and issue fees, we found costs to be between $3,000 and $9,000. More complex inventions will cost more than simple ones, and better attorneys will charge more than others. While these costs might not be a lot of money if you happen to be some giant like Google, for most of us even $3,000 is a meaningful amount. Especially when coupled with the maintenance fee you'll need to pay for a patent after 3.5 years, 7.5 years, and 11.5 years to keep it covered. Also, even if you have an amazing amount of creativity and problem solving capability, not every idea you come up with will be financially viable. It may be that there just aren't enough people who want the innovation, or the product that you are proposing is just too advanced for its time, or any one of several other factors.

That doesn't mean that you won't be pushed to hire someone to write up a patent anyway. Remember, lawyers do law. They don't necessarily do innovating. They don't necessarily see what happens in the boardroom with investors or out on the road at trade shows. They don't necessarily know how the rest of the process works. They probably will never suggest that you don't spend the money to get a patent and that you should just hurry up and blast your product out on the market before anyone else has the chance to catch up, avoiding all the upfront expenses. They usually don't advise you to do the patent work yourself. I've never seen a lawyer say to a client "that's a terrible idea." It's not in their best interests, and it's not what they're there for. They're a human trying to make a living just like the rest of us, so even when you go to a patent attorney with a crazy idea for a new type of napkin ring, which would probably be a terrible waste of money, you won't be turned away. But you should be. You really should. Someone should tell you the truth. There should be a service that helps you to evaluate your idea for its commercial worth—someone who doesn't have a vested interest in you spending lots of money. For now, you have to be honest with yourself because a patent attorney has no reason to do that for you. The point is this: When considering a patent attorney, know what you're getting yourself into.

Creating intellectual property and patenting it is a numbers game. If you're incredibly lucky you will be able to have one great idea and that will be the one that really pays off, but you are more likely to benefit if you are able to protect many ideas at the same time and get them on the market fast to see which one of them really makes it. In this case, you may benefit from a strategy that allows more creativity and more chances to protect more great ideas. That strategy can be filing your own patents and performing all the work that would normally be done by a patent attorney. This is called filing "pro se." A patent attorney will probably do a better job in the long run, but if you happen to have ten great ideas in one year those

---

[1] A non-provisional patent application requests the United States Patent and Trademark Office (USPTO) to issue a utility patent. This type of patent protects intellectual property rights for anything novel, useful, and non-obvious

services will cost too much. The question is what is the patent really worth and is it entirely necessary to have a "good" one? I have a partial answer that comes from my experience.

These days, in order to get most products out on the market and get manufacturing and supply lines open for a product, there is a lot of expense. In many cases, if you don't happen to have an extra $300,000 or more you will need to bring in some investors or take a loan. In general, banks won't lend on the basis of one invention. They just don't have the capacity to evaluate the worth or probability of success of a product idea, so a venture capitalist or some sort of angel investor may be necessary. In many cases one of the first questions that you will be asked is, "Do you have a patent?" Your answer has to be yes. The patent is invaluable then, but it doesn't always have to be an extremely well-written patent. It just has to be there to check the proverbial box.

Another compelling use for a patent is when someone is trying to come out with a product just like yours. In this case that patent will often stop them from even trying. Odds are that when they do a search and they see a patent with a title and drawings that truly expresses the idea, they probably won't pursue it. In both cases, the patent doesn't have to be an extremely well-written one; it just has to let you check the box. So that makes two very important, perhaps the most important, functions of a patent that you get as long as you simply have it.

Then there's another use for a patent in which that $3,000–$9,000 patent attorney comes in handy. Patents enable you to take legal action. But the truth is, if you are a private inventor, the odds are that once you file a cease and desist order to an entity that seems to be selling your product idea, you may not have enough money to mount a serious lawsuit anyway. The patent doesn't really protect you from someone copying and selling your idea, it just gives you the right to take anyone you think is violating your patent to Federal Court. Luckily I have never done that as I'm sure it would be quite expensive. As a private inventor this may be a bridge too far, so without proper resources the patent is useless to you in this situation.

Taking all this into account, the question is: Can you learn enough about how to write the claims of a patent, which is the most important and challenging part, to give you 99 percent of the value you get out of having a patent? You probably are. It's just a matter of study and some serious work.

For those that want to patent themselves, the basic steps are as follows:

- Conduct a thorough patent search
- Prepare the patent documents
- Register so you can file online
- Figure out how you will pay
- Submit the patent
- Pay the maintenance fees

It looks simple, and it is at heart, but there are some intricacies you'll want to be familiar with before you get started.

PATENT ATTORNEYS

We did some legwork to check up on our claims, corresponding with five independent patent attorneys as if we were inventors looking to patent a particularly poorly thought-out and poorly designed invention: a clip-on glasses fan designed to blow debris away from the lenses as shown below.

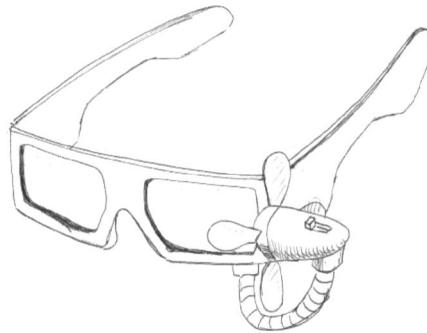

Pretty bad, right? Even after a few seconds of consideration, a number of serious design flaws are apparent. First of all, it's an uncovered fan that's blowing right next to your face, so if anyone with long hair chooses to wear it, they're playing a dangerous game. Imagine getting your hair tangled around something like this. Then, since it's attached to a regular pair of glasses, it'll have to be made really, really lightweight so that it doesn't weigh the glasses down and drag them off the bridge of the user's nose. That's not necessarily impossible, but it'd probably be very expensive to manufacture. The location of the clip itself is frankly ridiculous. It obstructs the user's view and could even damage the lenses. Not to mention that the entire contraption will certainly fail to achieve its purpose anyway. No fan that small is going to keep dust away for long, and it'll be blowing in your eye the entire time.

Regardless, as we talked with the attorneys, they responded as one might expect. They were enthusiastic about helping patent the idea, they were typically very upfront about quotes for patent searches and the like, and they never gave a straight answer as to their opinion on the device itself. When I asked for their opinion on the viability of the device itself, two of them said something along the lines of "we're just lawyers" as the reason they couldn't give any advice or assessments of the product's marketability. All of them mentioned in some capacity that they couldn't give an assessment of commercial success

One mentioned that he would never really say "no" to the product, even if initial searches turned up prior art that would make it un-patentable. Instead, he would take the time to work with the inventor through various redesigns until he thought the idea was potentially patentable. He also warned against invention promotion firms that would provide claims on marketability as those are typically scams. Another said he would provide an opinion letter after the initial search on whether or not the design was patentable, and that he would tell me a flat out no if necessary, saving me (the customer) some money. Of course, it's still in his interests to err on the side of positivity rather than to say no. One attorney put forth a statistic saying the probability of getting a utility patent[2] approved was 50% based on the USPTO's process and rather bizarrely recommended we get a design patent[3] instead. His reasoning was that they are easier to obtain and less expensive than getting a full utility patent, but failed to give an explanation as to why he didn't simply recommend a provisional patent. More on what all these terminologies mean will be addressed later, but the point is the suggestion appeared to be a complete waste of money. One has to wonder if this recommendation was based on the fees he could charge on that initial design patent application. Interestingly, in its literature about invention promotion scams, the FTC mentions avoiding those who encourage you to apply for a design patent because they tend not to be very applicable for most inventions.

The gist of what most said was that it was up to the inventor to determine commercial viability for themselves. Some recommended checking out crowdsourcing or Kickstarter and its competitors as a good place to start. Of course, that would be after getting a provisional patent. As they told me, it was essential to get out there and pound the pavement to find out for myself.

All of this is not to say that you should necessarily rule out using a patent attorney. Most of them tended to give good advice and were very upfront about what they could and could not assess. The takeaway is this: Just keep your eyes open when considering hiring one. Certainly if you want to be able to defend your patent against legal infringement and have a really solid document it may be worth your time, particularly if you truly believe your idea is a real money-maker. Just remember the attorney's role: to get you a legally sound patent, not to assess your invention nor market it. It really makes little difference to them whether or not you succeed; your fees aren't going to be any different, and you'll likely be done using their services at that point. Like they said, they are lawyers. They do law, not marketing, not inventing. That's still your job.

INVENTION PROMOTION FIRMS

There are some companies that may still make the kinds of promises and claims those attorneys wisely didn't. These so-called Invention Promotion Firms promise to help you through the whole process from obtaining a patent to getting your product in front of manufacturers and buyers. They may advertise everywhere: in newspapers, over radio, on TV, and of course online. Be extremely wary of them. Scams perpetrated by fraudulent firms of this kind happen with concerning frequency and can land people in debt while their innovative idea wastes away, bogged down by the firm's failure to deliver on what it promises. These scams have become such a problem that the issue has been addressed by legislation. The American Inventors Protection Act of 1999 requires a company to disclose a variety of information such as how many of their customers have achieved net financial profit from their invention before you enter into a contracted agreement with them. Even so, scams are still abundant.

---

[2] A **utility patent** is a **patent** that covers the creation of a new or improved — and useful — product, process or machine.
[3] In the United States, a **design patent** is a form of legal protection granted to the ornamental **design** of a functional item. **Design patents** are a type of industrial **design** right.

Companies like these promote themselves by making unreasonable promises, guaranteeing the patentability and marketability of an idea without acknowledging the high possibility of failure. They will frequently claim that your idea is one of the few that has real potential. Really, they tell that to every inventor. And why wouldn't they? The more they can pump up your design and promise results, the more they can get you to shell out for their services.

The FTC has a list of ways that people can identify the types of scams these firms perpetrate. They frequently offer a free initial kit or preliminary review to hook customers. They might then claim to do market research for the idea for a couple hundred dollars, but instead fabricate a positive projection based on nothing. Some companies are hard to reach even after multiple calls or might provide agreements over the phone rather than in writing. If they provide any sort of guarantee that your invention will be successful or patentable, they are most likely scamming you; there is never a 100% guarantee for either of those things.

A recent colorful example of such fraud was brought up in 2017 against a company called World Patent Marketing. The FTC brought claims against them for not delivering on their promised services. To add insult to injury, when customers threatened to report their unscrupulous behavior, the company reportedly sent them threatening emails promising legal action and boasted about "a security detail of ex-Israeli soldiers" more inclined to violence than conversation.

We've done a bit of legwork in this area as well. With a quick web search and a bit of investigation, we noticed that two of the first promotion firms that popped up had open lawsuits against them for fraudulent claims. We proceeded to contact some of the big invention promotion firms with our glasses fan invention to gauge their reactions. The representative from the first one we contacted, when pressed for an assessment of the commercial potential of the device, explained that he would simply inform me if the design was "something [they] can work with." Based on the initial information we sent in, he said it was. He, like the attorneys, said that the firm couldn't guarantee market success, though he remained encouraging. He mentioned that the "tool industry" is big and they would have a lot of manufacturers they would be able to recommend for our device. Overall, the firm was careful not to give any guarantees, but they certainly weren't discouraging. This is reflected in the disclaimers presented on their website as well, which repeatedly say that they do not give guarantees or assessments of marketability. This is something that can be seen on a lot of websites for these companies in an attempt to avoid claims against them.

Another of the companies we contacted claimed to be different from an Invention Promotion Firm. Their representative mentioned a few of the firms and warned us that those were typically scams. It wasn't really clear from their explanation exactly how this particular company was any different except that they only collect fees rather than a portion of the royalties from any agreements made on the device. I'm sure that statement was meant to be reassuring, but it really just means that they have absolutely no financial interest in the success of the invention. In fact, according to the FTC, "reputable licensing agents usually don't rely on large advance fees. Rather, they depend on royalties from the successful licensing of client inventions." So, you know, the exact opposite of what this firm was telling us.

Unlike the others, this company actually promised a free assessment of marketability on the product. They also claim to be extremely selective about the inventions they decide to promote. They were the only ones that, in the course of our communication, actually asked critical questions about the design such as how we would make it light enough and how we would prevent the irritation that might be caused by the product blowing in one's eyes. All of these things were tailored to make it appear as though they were not like the "others," but instead they just raised a number of red flags based on the warnings from the FTC and USPTO.

We went through with their free assessment and heard back from them the next day. Our funky little fan passed each category with flying colors and was wholeheartedly approved. In their assessment, they provided vague but encouraging assessments as to the device's patentability, marketability, and manufacturing capability, giving our invention a "high" rating for each category. They seemed to completely ignore the questions they'd asked us over the phone about the pitfalls regarding the design and insisted everything was well done.

They determined patentability based on six patents they dredged up, only five of which they attached for us. One was for goggles with tear-off lenses, one for a lens curing method, another for telecommunications enabled glasses, and two for lens cleaning apparatus: a spray foam and a lens washing machine. It paints a pretty clear picture of their assessment's thoroughness when the search we did, which took all of five minutes, turned up at least three patents that were much closer to our idea than these, including lens defogging fans and fans to extract heat from goggles. And yet, they concluded in their patentability assessment that "it is as clear as can be. Very positive." On the subject of marketability, they said only that "the market is very large for anyone wearing glasses of any kind. Very obvious." The final manufacturability section, typos and all, told us nothing beyond the fact that the "concept" could be manufactured with today's technology. Extremely selective indeed. Common sense and five minutes of research told us more about our idea than this supposedly rigorous assessment.

In general, it's clear these firms are not your friend. They may sound like it over the phone, with their encouraging words and warnings about the "other guys" who won't be so honest, but that's how they know they can hook people. Steer clear of their too good to be true promises and the neatly packaged fabrication of an easy, convenient path to commercial success. The truth of invention is not so rosy. It'll take a lot of hands-on work and effort to really get you out there. It's hard work, but with any luck it'll really pay off. Even if it doesn't, at least you know you did all you could. Don't let your idea rot because of someone else's empty promises.

THE USPTO

The United States Patent and Trademark Office is a federal agency of the US government. They are the ones with the ability to register and award patents on US soil, and they're the ones you'll be submitting to when applying for a patent. You can find them on their website at www.uspto.gov where you can find more information and apply for patents electronically. The USPTO accepts printed applications, too, though such applications can require additional filing fees that can really add up. Then again, who uses paper in this day and age?

Any information that you will need when filing pro se can likely be found on this site. The USPTO uses ESF-Web (an Electronic Filing System) to authenticate submissions over the web. The website may require you to download a Java script to start the authentication and application process should you want to file as a registered user, but other than that you should be ready to go. What you will need when submitting will vary depending on the patent. Later in this book you'll find a step-by-step example patent process, but if you want to go ahead and familiarize yourself with these features, take a look at their website, pro se filing information, and the patent search feature, which can provide you with an endless supply of patent examples.

PATENT SEARCH

A patent search is what ensures the novelty of an invention that you intend to patent. The USPTO will perform such a search when they review your application to ensure that there isn't already a patent or other prior art that encompasses it. However, you'll want to perform one yourself anyway so that you don't end up submitting for a patent that will only be rejected later. Initial patent searches can be done easily enough as an online search to see if something similar pops up, but a more thorough search requires more steps.

Start by defining your product or process. Come up with a list of all the terms that might come up when describing your product. Key words, related subjects, and whatnot can all bring up patented products that are similar to the one that you are trying to patent. Type these into your favorite search engine and see what comes up. Maybe your idea will pop up immediately as an existing product. Maybe you'll find something similar but the design is not the same as the one you had in mind. Maybe it doesn't come up at all. In any case, this will give you a good basis for your patent search. Now you can go deeper.

Google Patents and PatFT (found on the USPTO website) are good resources for finding any actual patents that have been filed. They can also be used to filter out patents in the field your proposed product pertains to. A lot of categories and fields have a CPC (Cooperative Patent Classification) code that can help you along the way to bring up these similar patents. These codes can be found by a simple term search on the USPTO website, though it may take some investigation to find the code that'll give you the category your idea fits to. After looking through the patents your code brings up, you'll also want to search your brainstormed keywords separately just in case. Then, if you want to be really thorough, you can search for foreign patents in your patent's field. If after all of this still nothing comes up that claims the idea you have in mind, there's a good chance that your design is in fact novel. Do keep in mind though that even if a product doesn't come up in your search that doesn't necessarily mean that there aren't others already in the process of developing it, so don't dawdle.

PATENTS, PATENT TYPES, AND FEES

There are three types of patents which you can apply for from the USPTO: Utility, Design, and Plant patents. Utility patents are the most common of the three, and most likely this is the one you will be applying for with your new product. Utility patents cover just about everything: a process, a bit of material, a machined part, or even an innovative improvement to something already being manufactured. Design patents are a little less broad, covering a new design for an existing product, and are often focused on the visual components of the product. Plant patents are a bit of a niche subject involving asexually produced plants. We won't go into detail about plant patents in this book.

In addition to these three official patents, you can apply for a provisional patent, usually with the intention of later applying for a non-provisional utility patent. Provisional patents act much like a reservation of a patent for a product still in production that will be filed for as a non-provisional patent on a later date. The non-provisional patent can then take the submission date of the provisional patents' submission. Provisional patents are often what people are referring to when they say "patent pending."

UTILITY PATENT

Utility patents are filed to protect the function of a product or process. In other words, this type of patent protects what the purpose of the product, process, etc. is. Having a utility patent means that you have full control over your product. For example, consider George Schneider, the inventor of the toaster and holder of the utility patent for toasters. Essentially, for the 20-year duration that his patent was in action, Schneider was the master of toasters. Any company, entity, or manufacturer who wanted to make a toaster had to go through him first. Likewise, owning the utility patent of whatever you've invented gives you full rights and ownership of that product's manufacturing.

The initial cost for filing a utility patent is $75 for micro entities, which is the classification under which most new inventors will fall under. To be precise, a micro entity includes individuals with an income under three times the national average who have fewer than four patents to their name. However, in addition to the initial filing fee, you will be expected to pay a search fee and examination fee which are expected to cost $165 and $190 respectively for micro entities. Once the patent has been accepted by the office, you will need to pay a $250 (for micro entity) issue fee and then your product will legally become your intellectual property. However, additional maintenance fees are charged 3.5, 7.5, and 11.5 years into the life of the patent, which for micro entities comes out to be $400, $900, and $1850 respectively. The utility patent will protect your invention for up to 20 years assuming that it is maintained.

You can file for a utility patent on the USPTO's website (www.uspto.gov). In addition to the fees, you'll need to submit the application itself with an abstract, industrial application, drawings, descriptions of drawings, and claims for the invention. Applicants are also required to submit three forms: a utility patent application form, an application data sheet, and an executed oath/declaration for each inventor.

In your application, the abstract acts as a summary of the overall application and gives the reviewer an overview of the content that will follow. These are ideally short, not exceeding 150 words. In-depth details and explanations should be saved for later in the application.

The industrial application is what lets the reviewer know what your invention is, what it is for, and how it may be defined or identified when seen on the market. You may also need to cite prior art in order to defend the novelty of the invention or else help examiners get a better grasp on what exactly your invention is and how it differs from what existed before. Prior art refers to any material relating to your design that was made available to the public before the patent was submitted. If your product is something already seen in prior art examples, then the idea is no longer considered novel and will be rejected. Prior art can include previous patent submissions, web articles, videos, and more.

Drawings are used to illustrate the appearance of the invention, show some key features to identify the invention, and more. These pictures will need descriptions of what they depict along with the full written description of the invention, again to give the reviewer a better understanding and definition of your invention. These will need to be clear, in black and white coloring, and labeled appropriately. If you made your invention in a CAD software, then you should already have everything that you need to make these drawings. It's important that you have enough drawings to give a clear idea of your product. It is recommended that you have several views so that an examiner has a full understanding of what the product is and doesn't miss any aspects. It can be a good idea to number these drawings as well to clarify the description and the features it pertains to. If there is a hollowed out section, or multiple layers to the design, it may be a good idea to have a sectioned view of the product that cuts through the center to give a better idea of what the inside of it might look like.

Finally, claims[4] describe exactly what you are taking ownership of with the patent application. They are by far the most important part of the application. You may have one or several claims depending on the complexity of the invention and how well it needs to be protected. Claims should be clearly labeled so that it's easy to see them for what they are, and should be numbered and divided into subsections when applicable.

As for the forms, the utility patent application form requires the inventors' names, the product title, and other pertinent information. The application data sheet is a separate form that includes bibliographical data pertaining to the patent. Finally, the oath/declaration is an official statement made by the inventor that includes the invention title and inventor's signature. One of these documents needs to be filled out for each inventor involved in the patent. All of these documents can be downloaded from the USPTO website.

PROCESS PATENTS AND PATENTING SOFTWARE

A process patent is a form of utility patent, mainly involved with a new process of manufacturing or the like that creates a novel outcome of some kind. In order for a process to be considered eligible to be patented, a test known as the Machine-or-Transformation test is often applied. While it is no longer the sole method for determining a process' eligibility, it is still the one that is most frequently used. According to this test, a process can't be classified as novel unless it is tied to an existing machine or apparatus, or unless it transforms a particular article in some way. A process you intend to patent can't be something you'd see in nature, nor can it be some completely abstract thing.

One form of process that can be patented is software. Under copyright law, the actual code to any software you develop is already protected the instant you type it out on a computer. No one else can

---

[4] In a **patent** or **patent** application, the **claims** define, in technical terms, the extent, i.e. the scope, of the protection conferred by a **patent**, or the protection sought in a **patent** application.

copy the same code without risking legal repercussions. However, the intended function of the software is not automatically protected. While others can't use, line for line, the same code that you wrote for, say, an app that tracks squirrel migration, there is no automatic protection to stop them from developing a squirrel tracker app of their own. This is where a patent would come in.

Processes and software can be covered under a utility patent and are applied for in the same manner as any other utility patent. The only things that may differ are the contents of the patent information, with less emphasis placed on drawings, and more on descriptions, tables, flowcharts, and the like.

CONSIDER A PROVISIONAL PATENT

A full-fledged patent can cost several thousand dollars, especially when attorneys are involved. Even then, when you apply it will probably take you years to get it. If you patent each and every one of your ideas before you make any money, you'll have to have some very deep pockets. For this reason, you may consider getting a provisional patent instead. The provisional patent protects you for a period of 12 months. During that period, you'll be able to develop the idea and sell it off before you have to spend the big bucks to get it to the greater market. If you get the chance to sell an idea at an early stage, you may not get the full financial benefit of your idea, but it can be a good strategy if you are not in a position to get funding.

Furthermore, provisional patents are not made public unless the applied-for date is otherwise referenced in a non-provisional patent you file later, so even if you choose to not go through with the non-provisional patenting process, or otherwise miss the 12 month cutoff of the provisional patent, there is no risk of designs being stolen from that patent. However, no extensions can be made to provisional patents past the 12 month cutoff date, so you'll either need to file a non-provisional patent, or you will lose the earlier filing date. Filing for a provisional application costs a $70 filing fee for micro entities. Because provisional patents are not registered or otherwise processed, no additional fees should apply.

In addition to the fee for applying, provisional patents need a cover letter with the names of all inventors of the product or process, the title of the invention, and other pertinent information. The cover letter form can be downloaded off the USPTO website. Other than that, information added to a provisional patent is much in the same as that required in a utility patent. Technically, a provisional patent doesn't require an abstract or claim, but it can't hurt to include either.

DESIGN PATENT

Design patents are filed to protect the visual aspect or aesthetic of a product. If your product is an iteration of a functional object, if it has a certain shape, or is designed to look a certain way, then this is the patenting type that will cover it. Design patents are much easier and cheaper to get than utility patents, but will give you less power over the invention than a utility patent would. Returning to the toaster example, let's say you wanted to create a toaster in the shape of a squirrel that when the toast pops out, it looks like a bushy tail. By getting a design patent, you now have control of that designed toaster squirrel. No one else can make a toaster that looks like yours as this would mean copying the design. However, that doesn't stop them from designing a toaster that is in the shape of a raccoon instead since it is the look you are protecting with a design patent, not the invention of a toaster.

Filing for a design patent requires a preamble, cross references to similar products, drawings or photographs of the product, a description of the product, description of its features, a claim, and an execution oath or declaration.

Included in the preamble are the names of the applicants, the title of the design, and a brief description of the design. The title is what is used to help reviewers of the application define the design during their patent search, so descriptiveness is key. The title should be able to articulate what the article is (e.g. "toaster" if we're using the squirrel toaster example).

Drawings of the patented design are very important to articulate how the design should look since that is what this patent is protecting. Make sure that you have enough views in your drawing to encompass all aspects of the design, and include figure descriptions explaining what view each picture shows. While drawings can be clearer, black and white photographs of the design will also work, but it has to be one or the other. You can't have a mix of pictures and drawings. Colored pictures are accepted only when necessary and only after filing a petition explaining their necessity. Unlike a utility patent, a design patent can only have one claim describing the design to be patented and is often times described as "an ornamental design of X" where X is say, a toaster or something.

Like a utility patent, design patents require an initial fee, search fee, and examination fee. For a micro entity, these come out to be $50, $40, and $150 respectively. Design patents do not require you to pay any additional maintenance fees, but do require a $175 (for micro entity) issuing fee, and they only last for 15 years.

PATENT FILING EXAMPLE

This section will walk you through the patenting process. Of course, the whole thing starts with an idea. Just the other day I was in a brand-new plane; I don't remember the airline, and to my amazement the toilet seat had a little tab on the right-hand side that was meant for you to lift with your foot. I'd been thinking of this as an after-market device for years. Also, when it was time to put the seat back down, you didn't have to hold it, you just gave it a nudge and it slowly lowered due to a clutch device in the hinge. This was another thing that I had been thinking about for years but never put the time in to see if people would like it. These little things made the experience so much better. I was very impressed.

Why doesn't every toilet in the world have a little kick tab? I'm not sure of the answer. Is it impolite to use your foot? Many people do it anyway, it's just not so easy. Here is the CAD for an after-market device to accomplish what that airline did. It's a simple plastic tab that comes with pressure sensitive adhesive and/or stainless-steel wood screws. You peel off the tape and you press on the tab. Or, if you feel more adventurous, you can drill two little pilot holes and screw it on. It's that simple. A million-dollar idea? Who knows. Is it patented? Let's see.

FIGURE 79

Our first step to scour the prior art can be a simple web search of the words "toilet tab." Looking it up already turns up a product consisting of a tab with an adhesive peel-and stick attachment. You might think our idea is dead in the water now, but we're not finished yet! Let's look a little further to see if the idea has been patented and what the claims are. To do this, we must conduct a patent search using the processes explained in the earlier section to see if there are any pre-existing patents that utilize this, or a similar patent design.

In this case, our design is a handle that we can use to lift a toilet seat. Terms that come to mind include toilet seat lifter, toilet seat tab, toilet seat handle, sanitary handle, adhesive connector, etc. Another way to find related patents is using a CPC patent classification code. This too will take a bit of searching using the USPTO search engine. In this case, I found A47k13/10 to be the most applicable. A47k13/10 includes all devices that raise, tilt, or lower the seat or cover. The code is defined by its several classifications: "A" involving human necessities, "A47" specifying furniture and that sort, "A47k" specifying sanitary equipment and toilet accessories and the like, "A47k13/00" specifying relations to seats or covers and "A47k13/10" clarifying the intention to lift said seat or cover. As lifting a toilet seat is our intention, this description could not be more on the nose.

Now, using the USPTO databases PatFT, and AppFT we can look up our brainstormed terms and CPC code, and use them to systematically bring up all the related patents, taking note of the ones that are particularly similar to our design in a separate file with the intention of reading them over in more detail later. This search brings up a patent that looks a lot like the one we found on our simple web search online.

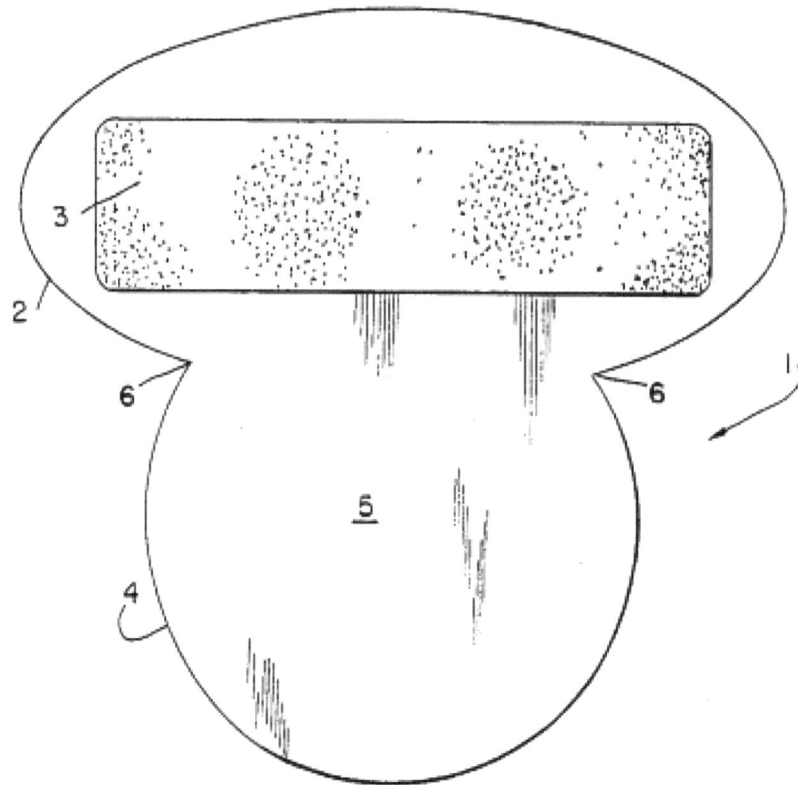

FIGURE 80

This patent, in layman's terms, claims the design of a flat, rigid, plastic tab that has a surface acting as a handle, and one acting as a mount to a toilet seat. However, more specifically and pertaining to our situation, it claims a toilet tab which is mounted to the underside of the seat using a sticky adhesive tape. Additional results from the patent search showed designs for similar patents which instead utilized screws to mount their respective handles and tabs. This means that our current design will not work as a new and novel idea if it is utilizing a sticky adhesive or drilling for a mount. It's time for a redesign.

Who needs sticky tape and screws anyway? These things may work if you want a permanent application of a tab, but what if you wanted one that is removable or readjust able? Adhesive is a no go, it'll leave sticky residue on the seat, and then you'd need to apply more tape. Screws present a similar problem; you can't move that handle around without leaving dozens of holes all over the seat. Our new design takes these issues into account. This one uses a clamp and screws to attach it to the seat.

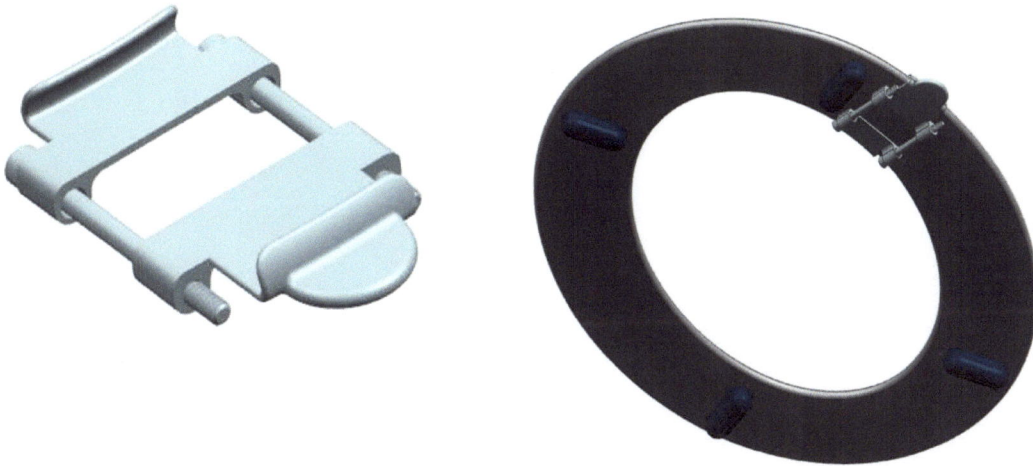

FIGURE 81

Now that we have a new design, there are more details that we should be adding to our search terms including clamp and clip. In the end, it was a good thing that we did check, because once again it looks like someone again beat us to the punch.

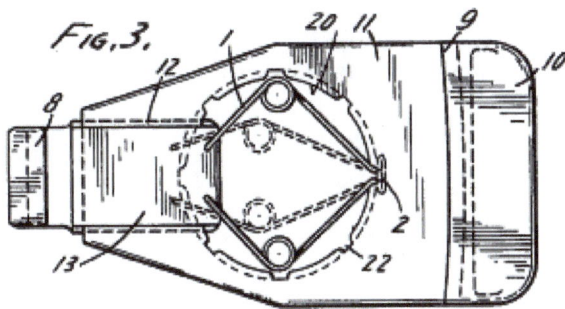

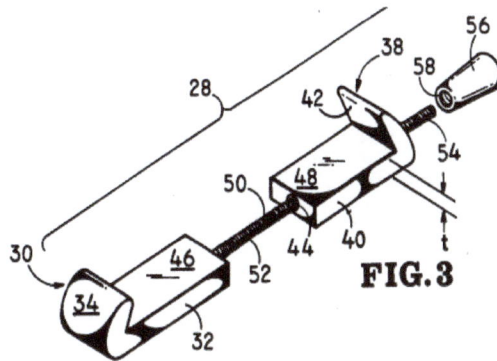

FIGURE 82

Both patents involve the use of a clamp or clip-like attachment to affix them to the seat. The one on the left even uses the same screw tightening clamp that we came up with. It's back to the drawing board again! Remember, everything that exists can be improved in some way. So we can't do our original idea for a clamp, but maybe there is a different way in which we can design the clamp mechanism. Say, springs, for instance. With a torsion spring and some clamps we could easily set up a seat tab clamp system like the following picture.

FIGURE 83

Another quick search, adding spring, spring powered, and torsion to the list, yields no existing patent or prior art with the same idea. So now we have a novel idea which we can submit to be patented.

We'll apply for a provisional patent with this design. When writing the application for a provisional patent, it is important to be clear and concise. Any information provided will be useless if patent examiners cannot discern the nature, intention, and novelty of the product when you actually file for a full utility patent. As we've mentioned previously, a provisional patent will require a description of the patented product, drawings and sketches to better illustrate design intent, and a cover letter that lists all the named inventors and other such information. Remember, a provisional patent doesn't need to be approved by an examiner so things don't need to be quite as detailed as the non-provisional version will be. Even so, it doesn't hurt to be thorough.

We downloaded the provisional patent application cover sheet on the USPTO website. After filling in our name, invention title, customer number (you can enter a mailing address if not registered as a verified user), entity status, and signature, we were ready to go. We then converted the files into PDF format and started the filing process. You can file with the ESF-Web service either as a registered or unregistered user. Registering as a user will require some work but can ensure there aren't any mix ups with personal information getting confused down the line. It also can make the patenting process far more streamline down the way.

To register, we needed first to fax a filled out copy of a Customer Number Request Form that we found on the site to the number on the website (571-273-0177 at the time of this writing).

PTO/SB/125A (11-08)
Approved for use through 03/31/2021. OMB 0651-0035
U.S. Patent and Trademark Office, U.S. DEPARTMENT OF COMMERCE
Under the Paperwork Reduction Act of 1995, no persons are required to respond to a collection of information unless it displays a valid OMB control number.

| Request for Customer Number | Address to: Mail Stop CN Commissioner for Patents P.O. Box 1450 Alexandria, VA 22313-1450 |

FIGURE 84

The Customer Number takes around a week or so to process, and then when it is, the USPTO will send it to you via email and snail mail. After we got our customer number, we were then able to fill out a Digital Certificate Action Form and had it notarized and mailed to the specified address.

PTO-2042 (07-2009)
Approved for use through 11/30/2014. OMB 0651-0045
U.S. Patent and Trademark Office; U.S. DEPARTMENT OF COMMERCE
Under the Paperwork Reduction Act of 1995, no persons are required to respond to a collection of information unless it displays a valid OMB control number.

| Certificate Action Form | Address to: Mail Stop EBC Commissioner for Patents P.O. Box 1450 Alexandria, VA 22313-1450 | USPTO Use Only |

FIGURE 85

Getting this form filed can take a while, and again we had to wait a week or so before it was processed. When the action form is processed, the USPTO will send two codes, a reference number and activation code which will be sent in two separate emails. We used the link provided in the email sent with our reference number which we could use to proceed and get our Digital Certificate File. Make sure that if you get one of these, you save it in a place you can easily find. When everything was processed, we were then able to authenticate our account and become a registered user. Again, a bit of a time-consuming process, and perhaps a bit unnecessary if you only ever plan on filing one and only one patent, but if you plan to file more it makes things much simpler in the long run.

When you are ready to file for a patent, you can do so by signing into MyUSPTO.

FIGURE 86

In your MyUSPTO account, you can file patents under the tab Patents under the name File Patents.

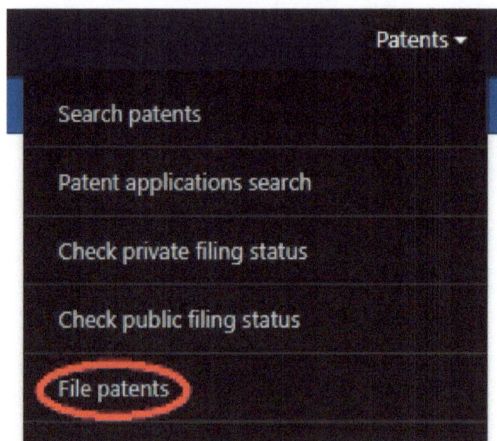

FIGURE 87

As a registered user, you can file patent using your digital certificate. Again, you can file patents as an unregistered user by imputing your personal info instead. Filing as a registered user will require you to have downloaded Java Runtime Environment. (Don't worry, it's free.) Using the new method, download the authentication script and open it with Java Web Start Launcher (javaws). This may take a bit of searching through your computer files.

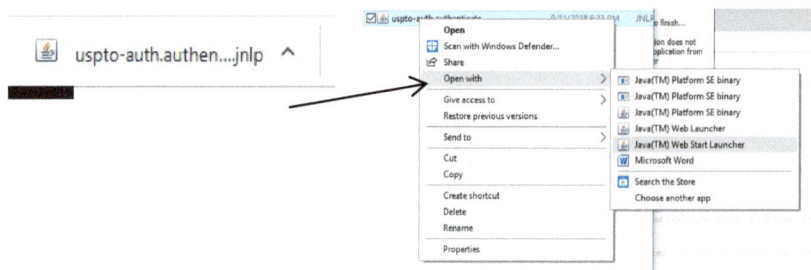

FIGURE 88

When all your information is included, or the authentication process is done, we can upload the three necessary files for the patent provisional (a cover letter, description, and drawings) labeled accordingly. When ESF-Web reviewed everything and found no errors in the submission, we were able to calculate our expected fees, submit, and pay. You are now the proud owner of your very own provisional patent. Keep in mind though that this is only a provisional patent that we applied for, a "patent pending" status, so additional filing will need to be done within a year should we decide that we want to go through with the non-provisional utility patent later on. In the meantime, you have the time to contact distributors, make contacts, and get the ball rolling on manufacturing and distributing your design.

FOREIGN FILING

Now if your patent is successfully filed and approved, you have protection for your product being sold in the US. But what if you want to sell elsewhere? You may then need to file a new patent application for those foreign countries. In order to file a patent in foreign countries, you can either file for each country separately, or you can file under the PCT (Patent Cooperation Treaty). Filing under the PCT will make your patent protection valid in all countries where the PCT applies.

The submissions for foreign applications will require the same, or similar, information as it would in a US patent, though when applying to an individual foreign country, you will need to have the patent's information translated into the official language of said country. The PCT filing only requires you to submit the application data in one language rather than the official languages for each country. The application for a PCT patent will require a 1330 CHF (Swiss franc) filing fee, which is about $1343.30, along with a 150-2000 CHF ($151.50–$2020) searching fee and a small transmittal fee (which varies).

WORKING WITH OTHERS

Some say "You're only as good as the partners and associates that you work with". You must find good people that can help you if you want to bring an exceptional component to market.

Getting your product to market is not a solitary endeavor, so by all means, find the right experts for the parts of an invention that you don't really know enough about. Few of us, if any, know everything we need to know about the entire process, from performing the basic patent work, to finite element analysis, manufacturing, marketing, packaging, and distribution. That's just too much for one person in one lifetime to know. Many inventors are extremely bright people who can figure out how to do almost anything. This is both a blessing and a curse. When you take the time to figure out how to do something new and difficult, it's a great feeling. On the other hand, it's been said that "those of us who act as our own lawyer have a fool for a client."

When we try to do all of our own work, we guarantee that the jobs can't be done in parallel. We can no longer leverage the skills of others, which means that our product gets to market later than it otherwise would have. This may mean death for a new product. Also, the effect of experience can have an exponential impact on your ability to perform a task. An experienced CAD jock can knock out a design in a tenth of the time it takes for a person who's just learning how to use the program.

For example, if your invention is a simple device used to help people lift large pieces of sheetrock, you can easily make a wooden prototype, but when it comes time to create a plastic design that can be injection molded, you really need someone who has experience defining injection mold geometry. You need someone to create an accurate CAD model for you. You need someone to work with the injection molding manufacturing firms who can tell if they are really producing the geometry correctly and someone to help you select the right material.

Then there's the distribution network for the product; you need someone to call buyers and write up contracts with those who would sell the product. Along the way you need a graphic designer to design the package, the box, and the little instruction sheet that goes with the product. Maybe you want a cartoonist to create a nice instruction sheet that can be understood by anyone in the world because the instructions are all in pictograms. Without enough money you may have to do all of this yourself, but chances are it's too much work for you to do alone.

In order to get a great product on the market and make a real profit from it, a lot of components have to come together. The timing must be right, the engineering has to be great, the manufacturing has to be optimal, and the marketing has to be appropriate if not innovative and groundbreaking. Consequently, you will probably need the services of many different people. It's great if these people have actual formal training, but I believe that it is far more important to select the people who have had experience and a proven track record or are at least incredibly visionary and successful.

It's great when you have all the components coming together beforehand. It's great if you've already been successful with a similar product or service, but in many cases, especially if you're trying to invent and promote a new product, you don't have all the pieces in place. Even when you do there's no guarantee; you can still be a victim of random chance. Set your product up for success by surrounding yourself with great people who will help you succeed.

FROM NEW TEAM TO WELL-OILED MACHINE

There is a common idea in the room of engineering about how every team is formed into an organized group. It's the idea of forming, storming, norming, and performing. In the forming stage, you are just getting started. You go out and find qualified members that you think have the qualifications that will work for you. You get the team all together and start working on whatever it is that your team has come together to do.

Working with a team, when it first gets made, can be difficult and people start to storm. Many members might be unfamiliar with the other members. Some might not know what their function in the team is, and others just can't find where they fit into place. It's a struggle as everyone tries to collaborate with each other while maybe not being on the same page, or otherwise not understanding what they need to do.

The norming stage is when all these things start to come into place. People start understanding what their function is, and team members start getting along better. This continues on through the team's track into success until they really start performing as a proper team. It's good to have a team like this, there's only so much one person can do on their own after all, and you may find that working with like-minded individuals is a surefire way to make long lasting relationships, friendships even, that follow you for the rest of your life.

BUILDING THE RIGHT TEAM

Now that you recognize the need to work with others to get your product out into the world, you'll need your team. It is not always possible to build your team from scratch, but when you have the opportunity to do so, it's great. You just need to know where to start. Perhaps the first thing you need is a way to find good team members. I'm sure an entire book can be written about the various ways of finding new employees and partners. At the time of this writing there are a huge number of websites and organizations that can be used to find these people. There are trade shows and all sorts of other ways of getting introduced to people who may (or may not) be right for the team. The "old boy" network is alive and well, too. The people on your swim team or your volunteer group or your church and your cooking club all know people that know people that can help you. In short, you gotta get out there.

But it's not enough to just have a bunch of skilled individuals. You can have great individual team members who are working against each other because the team is not configured well or their personalities clash. In a team, everything must be balanced. The wild-eyed dreamer must be balanced by the schedule-oriented pragmatist. The super amazing nerd genius must be balanced by the gregarious get-out-there-make-a-deal-and-grease-the-skids type. The incredible optimist drive-to-the-goal type needs balance from the loving pessimist.

Although it's a gross oversimplification, when thinking of the kind of people you'll need it can be useful to consider the following statement: There are two types of people, maintenance and visionary. Maintenance people have a great internal clock and are great at repetitive tasks. They're happy doing the same thing perfectly each time. They resist change, they're risk averse, and they don't miss a thing. They follow the rules and hate it when things are out of place. They're organized and reliable, but they can't come up with a new idea to save their lives. They get stuck in the same rut even if it doesn't work well. They don't always notice when the thing they're doing doesn't have a great result. When they're taught how to do something it's locked in, whether it works or not. They tend to be myopic. They may respect authority even when the authority is deeply flawed. They won't make waves. When these folks are asked to innovate, it's a train wreck. When they are taking care of a complex task that has to be done well, they're wonderful. If you have a fleet of 747s and you want to keep them in the sky you want a lot of maintenance people. But every now and then there's some really difficult problem to understand and solve, and you really need an out of the box fix. You need a visionary.

Visionaries are odd ducks. These folks love to ask the question "Isn't there a better way?" They have a lot of diverse hobbies, they don't stay on one thing, they're incessantly asking questions, they're critical and imaginative. Innovation is what they thrive on. If they're in a stagnant situation they won't last long. When they perform a complex task, they will first ask, How much of this is really necessary? They will pick and choose action items that are part of the task and do the ones that they think are necessary. They will assume that the fine details are unnecessary anyway, and they will leave it undone and go onto the next thing. This will work sometimes, and they will get a job done in a super-efficient way that truly saves resources. Other times they will leave a detail undone that they think is unimportant when in reality it is.

When it's hiring time, you have to ask the question, What does my company, or organization within that company, do? Is it a high tech company that is competing with other high tech companies in an industry where there's a new product out every month? Or is it a company whose nature is that of doing detailed operations that have to be done accurately? If you're running an optometrist's shop, you may be poorly served by the visionary. You definitely want the maintenance guy. In fact you'll want a bevy of maintenance folks. You'll want them to adhere to strict protocols, and follow every rule and procedure that is time tested and proven to benefit the customer. However, you will be in competition with other optometrists who are doing the same thing. If you want to have a better market share, you're going to have to hire at least one innovator.

Now you're faced with another question: How do you pick 'em? At Design Visionaries, when we interview we use a number of mechanisms to ensure that we get the right type of people on our team. We want to know that they are very good in their field; they are kind, communicative, positive, reasonable, rational, reliable and, most of all, loyal. That's way too much to get out of one interview. It takes several interviews and real testing.

Out of all the special methods we use to find employees, one of the most unusual is a simple task. We ask prospective applicants to produce a drawing of a bear sitting on a drafting table.

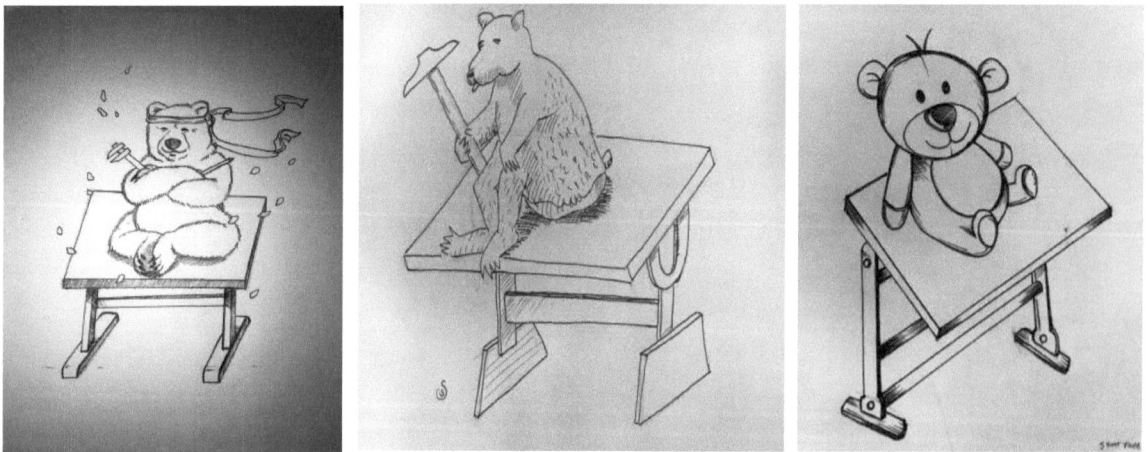

FIGURE 89

In this exercise, we're looking for a positive attitude, calmness under pressure, and creativity. We don't really care if the applicant can actually draw, although in the product design business that's a huge plus. Instead, we really want to see our applicants try to do something that is a bit difficult on the spot. We don't want to hear, "Oh, well, I'm not a good artist," or "I was never good at that sort of thing." We're looking for someone who will smile at the challenge of being asked to do something unusual. The product design business is full of extremely creative, bright, and dynamic people with odd hobbies like building go-carts or underwater camera photography. It's very exciting to find these individuals and get them on the team.

LEADERSHIP

Building the right team begins with getting the right people with the right skills, but it in no way ends there. When you have great team members, they can't produce well unless there is a great framework in which to work. You must have clear roles laid out for those people that will help you succeed. The roles that your team members have must dovetail so that the team works like a well-oiled machine. Communication must be good and frequent, and team members must respect each other's contributions. It is essential that there is an underlying philosophical framework that everyone acknowledges and is willing to put their energy into. They have to be able to trust that when they invest their time and/or money they have a great chance of getting a return. There needs to be good feedback among the other team members, an evenness of sacrifice, and a common general appreciation.

There are many books written about team building, and it's not practical in this text to duplicate what's already out there, but as an entrepreneur it would be unwise to try to build a team without putting a lot of thought and energy into all the groundwork. The most important groundwork that can be done above and beyond good, sound contracts, fair remuneration rates, great working conditions, and communication is to underscore the basic "personality" of the company.

It has been said that a company takes on the basic personality of the founders. In my own experience, this couldn't be more true. There's an old adage, "If you want a friend, be a friend." It goes right along with what's necessary to have a good company. If you want a good company in which people produce and do great things, you need good leadership. You need people who are willing to lead from the front. These people think about the business all the time because they love it and they love the people involved. They inspire folks to go beyond what they would normally do. It's great if people are inspired by leadership. It's really tough to work for someone who doesn't sacrifice and asks employees and team members to do things that they themselves would never do.

President Truman famously said "It is remarkable how much can be accomplished if you don't care who receives the credit." Truman lived up to this quotation in the case of the famous Marshall Plan. He and his administration knew that if Europe was allowed to remain devastated and impoverished for a long period of time after the close of World War II, it would be a breeding ground for all sorts of unrest that would eventually hurt everyone. To solve this issue, Secretary of State George Marshall along with various others came up with an extensive financial assistance plan to help Western Europe. One of the minds behind that plan, Clark Clifford, wanted the plan to be named after Truman, but Truman declined that suggestion. He knew Marshall deserved the credit, and knew his opposers in the House and Senate would never let something with his name be passed. So they gave the credit to George Marshall, the plan passed, and most economists agree that it was a huge success.

Give your team members the credit and appreciation they deserve. In the design world, credit is more than a name. It's financial prosperity, it's ownership, it's participation in the decision making and the branding. In many cases it's as simple as praise. A simple "Thank you, job well done" can be worth more than the big fat check. Of course, it's better to give both.

ORGANIZING A TEAM FOR SUCCESS

As I said before, having the luxury of building a team from scratch is rare. I believe most companies grow from small beginnings, then add on employees over time. The question then becomes, when you have a staff, assuming that you're not going to fire them all and start from scratch, how do you get that staff to be more innovative and more detail oriented at the same time? How can you strike a balance between maintenance and innovation in an organization? Naturally, the answers are many, but here are a few ideas.

First, on the innovation side, you must provide an atmosphere that encourages risk and accepts certain types of failure. This is very difficult because people often want to do everything "the right way." The innovative atmosphere must not be inflexible. It has to be light, communicative, and have its rules be the bare minimum.

Companies try to get their employees to achieve their innovative potential in a plethora of ways. One famous method comes from 3M with its "15% time." With this policy, the company encourages workers to use a portion of their time to pursue projects of their choosing, encouraging them to innovate and experiment with something that excites them. This time is credited with notable innovations like the ubiquitous Post-It Note and has been copied by many other companies in some form or another.

Some companies have a formal process that they use to ensure that everyone's creativity is as engaged as it can be. I applaud them. Other companies have a more streamlined process because they have such trust in one another that they don't need a formal process. I applaud them, too. It can work either way. Still other companies have one amazing nut of a person who seems to be far more creative than almost anyone. The genius comes up with most of the ideas and everyone else is there for idea refinement. This can work too as long as those who are in the supporting role are valued as well. It's important to note, though, that ideas come from people, not processes. Ideas can be facilitated by a good process that somehow encourages and supports the ideas, but in the end, having the right person or group looking at a new design and nourishing it is the most important task. That being said, there are some elements that are easy to recognize in a creative sort of atmosphere as well as some that are indicative of the opposite.

A Creative Atmosphere Has:

1. Opportunities to experiment
2. Broken stuff and old machines to take apart (teardowns)
3. Manufacturers coming in to give demos of their processes and capabilities
4. Space to get dirty without upsetting the gestalt of the place
5. Free and easy communication
6. Energy
7. Collaboration
8. Encouragement to try something new
9. Room for randomness
10. Many varied capabilities and tools at your fingertips
11. Miscellaneous material samples

An Uncreative Atmosphere Has:

1. Levels of management
2. Offices with doors that close
3. Communication by email
4. Schedules and meetings for the sake of schedules and meetings
5. Reports of progress instead of real progress
6. Extensive approval process with copious validation and review
7. Anger and harsh criticism
8. Compartmentalization

On the maintenance side there must be structure and schedule, planning and clear goals and rules. These things are the backbone of a successful company. Without these important mechanisms there will be chaos and waste. There will be otherwise perfectly good opportunities lost. There will be failure. These are attributes of an organized and well-maintained company. But a company that is set up for

A Maintenance Atmosphere Has:

1. Clear rules
2. Great records
3. Clear and well-defined agreements
4. Good planning with details and milestones and schedules
5. Neatness and attention to detail
6. Regimentation and "automatic" systems
7. Stability and clarity
8. Frugality and savings
9. Insurance

A Non-Maintenance Atmosphere Has:

1. Rules that are ignored
2. Different rules for different people
3. Inconsistency
4. Clutter and instability
5. Emotionality
6. Poor record keeping
7. Unanalyzed spending and investment

DO THE ANALYSIS

You must know every fine detail about your invention to a fare-thee-well. You must obtain a detailed definition for your invention. These days, that usually means a CAD model. This includes things like detailed drawings with tolerances and inspection data, material specs, test results, and everything else you will need to ensure that when your invention is manufactured, it will be safe, it will function as desired, it will be reliable, and it can be manufactured for the lowest cost. You must also think about scheduling, shipping, packaging, and the full ramifications of the distribution network that your invention will be associated with. In many cases you must take into account the Christmas season and, if you're getting injection molding in Asia, the Chinese New Year. In the end, knowing all the details is the only way of mitigating the various risks that your product poses to you and the user community at large.

WORKING WITH A GOOD, EXPERIENCED ELECTRICAL ENGINEER

When a new product design has some electronics in it, you need to consider using the services of an electrical engineer. Mechanical engineers work on just about anything that moves, but electrical engineering is very specialized. The electrical world is more hidden than the mechanical world. There are many people who have never been formally trained in mechanical engineering who can whip you up a great go-cart made out of a lawn mower. With electrical engineering, similar skills are very rare.

When you use the services of an electrical engineer, it's important to know what will be needed. First and foremost, you as the inventor and customer of the services must have a very good idea of what is to be built by creating a Product Requirements Document (PRD). (If you skipped the PRD section in this book on page 48, now might be a good time to take a look at it.) Once the PRD has been established, the electrical engineer can begin.

The first thing the electrical engineer will do is create a functional spec. This is yet another document that contains very technical information that is meant to satisfy the requirements of the PRD. The functional spec will include things like power requirements, the number of LEDs and their colors, the exact size of the video display, the amount of time that the product can go without being on the charger, and myriad other real-life considerations. As the functional spec is being created, it is hopefully being compared to the PRD constantly. It's no fun when the two activities are done in series, only to find that, at the end of the process, the PRD and the functional spec don't agree. Sometimes this is unavoidable because product designers may desire a combination of product attributes that are conflicting or impossible based on the budget. However, the two documents eventually have to be reconciled.

The next step is material procurement. The actual components have to be weighed and considered. The part numbers are found from catalogs with names like Digikey™. These components are fashioned into a working prototype. Somewhere in the process, the prototype is tested and tweaked based on the testing and a "schematic capture" can be performed. The schematic capture is the technical definition of everything that is being used and how it interacts, including the exact part numbers of all the chips, capacitors, and resistors. The schematic capture is necessary to create a board layout.

The board layout has the exact positions and sizes of all of the components. The board layout is used by the mechanical engineer, and in many cases the mechanical engineer and the electrical engineer have to work together so the mount holes line up and the ribs and other internal structures don't collide with any of the components. The tolerances of the board and the tolerances of the mechanical components surrounding the board may have to be considered to ensure a good fit. Occasionally, the mechanical engineer provides "keep out" areas to the electrical engineer in order to ensure that the components won't interfere.

Once the board layout is done, you'll need to choose a good PCB (Printed Circuit Board) manufacturing house to perform board fabrication. Naturally, a few test boards will be created before an assembly line can be put together to create thousands of copies. A good board manufacturer will perform tests on the board to ensure reliability as the boards are being manufactured.

MANUFACTURING

Once you have defined a great product there are many things to know about how to get it manufactured. Here are some items that will allow you to avoid some of the pit falls:

GET A BUNCH OF QUOTES

Once you have a design ready to be built, don't be afraid to get a bunch of people working on a price quote for manufacturing. Make sure they sign an NDA, and get them to work. Sometimes, when a supplier is busy, they will quote high because they will have to bend over backward to help you at that time. Because of this, the place that gave you a good quote before may not give you a good quote every time, which is why it can be a good idea to keep your options open.

Read through the quote carefully; you can learn a lot from it. For example, recently we were designing an electromechanical device that had a large mechanical actuator inside of it. We looked online for a lot of the components and found that the prices were all over the map. We also waded through a lot of quotes for the fabrication of many of the components; they, too, were all over the map. The design included some large sheet metal parts that seemed very expensive. By carefully reading through the various quotes and talking to the suppliers, we were able to get the overall price down quite a bit by making small changes to our order to cater to the abilities of certain manufacturers. We went with local folks that we could visit and have face to face conversations with easily. Even in this world of lightning fast communication and video conferencing, there's still no substitute for the face to face.

WHEN YOU COMMUNICATE WITH MANUFACTURING, COMMUNICATE WITH MANUFACTURING

It sounds odd, but many designers and engineers make a huge mistake when they expect that just because they have prepared a detailed drawing with geometric dimensioning and tolerancing and they have sent it to manufacturing, they have done their due diligence and their component will be manufactured correctly. In fact, some folks think of a detailed drawing as nothing more than a legal document that guarantees that when you get parts from the vendor, you can inspect them and then reject them if they don't conform to what you want. Although that is essentially true, it doesn't make for a good relationship between the designer and the manufacturer. In reality, a good working relationship with a manufacturer wins over in the end. If you allow the manufacturer to give their input early on in the design process, you will have a better understanding of what is possible in the manufacturing process, and you have a chance of a better or less expensive product.

When you allow early input, you can ensure that the product definition data you gave to the manufacturer is interpreted correctly, and you will have a smoother path to the end goal. When you talk to the manufacturer, you will understand that some of the information that you spent time putting on your detailed drawing will be ignored by the manufacturer or may even be misinterpreted. In some cases when manufacturers see the definition for the first time, they don't realize that there's a portion of the design that they can't in fact perform. If you are in good contact with them you can perform slight tweaks to the design so that they are able to manufacture the design properly.

For example, I recently designed a meter for one of my clients. The manufacturer that was chosen was located in China. As the design was delivered, I flew to China to be on site if they had any questions. The meter had a certain metal plate on it that had to be attached in a special way. When the manufacturing people began to really look at the design, they found that they couldn't produce the metal piece as it was designed. Luckily, I was there with a CAD system on a laptop to change the design to fit the ability of the manufacturing house. The fact that I was able to have a face-to-face communication with the manufacturer and make the quick change saved the day.

WORK WITH PEOPLE WHOSE COMMUNICATION STYLES YOU LIKE

Sometimes it doesn't matter how good a design or manufacturing supplier is—they just don't communicate in a way that you appreciate. Perhaps they are too curt or too wordy or you just can't understand how they do things. Maybe their version of honesty is not the same as yours. Perhaps they don't have an FTP (File Transportation Protocol) site to allow you to get them your files. Maybe they aren't good at getting back to you with emails or scheduled calls.

You need to be able to communicate well with the firm. Bad communication causes more problems than faulty calculations in most design-to-manufacturing processes. A lot of people in the industry like a detailed paper trail. Others like the face-to-face method. Both of these have their pros and cons, but the best method is probably the one that you're most comfortable with. There are a few things that are important to remember. It's important to communicate early and often about problems or perceived problems. Avoid emotional outbursts when getting your concerns across. Keep in mind the strengths and weaknesses of different forms of communication. Email can be easily misinterpreted because you can't hear their tone of voice so it's often better to call. Email is great because it leaves a nice paper trail to ensure communication is preserved. Texting is useful because of its immediacy, but it's usually not as detailed and formal as an email.

TEST YOUR PRODUCT

Your product must be safe. The worst thing that can happen to an inventor is his or her product accidentally hurting someone. Above and beyond the lawsuits and loss of money and wealth is the guilt you would feel if, say, a toy that you invented injured a child because of a miscalculation or a material that was supposed to be safe but turns out is not. Thankfully, there are people to help you in this regard. In the 1970s, a bill was signed that created the Occupational Safety and Health Administration (OSHA). That organization officially authorized a number of Nationally Recognized Testing Labs (NRTLs), the most prominent of which is Underwriters Laboratories Inc. A number of standards, tests, and certifications were created to ensure that product designers can make their components completely safe. When your product is compliant with the recognized standards, you can obtain a "UL" listing which means the labs will stand by your product. The UL listing certifies that your product meets the standards that are required by distributors that will get your product out on the shelves. Not every product is of a nature that it requires a UL listing, but certainly anything that will be plugged into a wall outlet needs one.

Once you have successfully gone through the UL listing process, you can display the UL logo on your product. If you want to sell your product in the EU, you must have a CE listing. There are other listings such as the CSA and the TUV that you'll find on many of the products that you already use and own. In order to obtain these, you will have to go through an independent lab. When you work with them, they will ask you a number of questions to find out which tests and certifications are appropriate for your product. This process can be very lengthy and very expensive. If you don't plan for it you will be surprised, and it can easily kill your product delivery schedule.

Even in more general circumstances, unexpected issues with testing can kill a product. A few years ago, I was contacted by a former client of mine who wanted me to help bring a product to market. He didn't have enough money to fund the development himself so, like many others, he wanted me to do the product design for a percentage of the company that would be created around the idea. I had been offered this deal many times before, and I generally declined. There're too many variables that would be out of my control if I partnered up with a total stranger just for a "great" product idea, but there was something about this time that I thought would give us a better chance of success.

We worked with a few other partners who I knew from previous experiences. The product idea was a power outlet with two built-in USB power ports. The partner already had the patent. The electrical engineering was basically done, all that was needed was the packaging. We decided to go for it, and we formed a limited liability company called Currentwerks. From the onset we had a good strategy. We intended to fund our development costs and a lot of our other activities through the sale of goods that we were producing along with other funds that we put together from the partners. We kept the expenses as low as we could. We didn't have a formal office and a lot of the expensive services that we would have had to pay for we just did ourselves. We did the design, testing, legal work, set up the manufacturing, packaging, and everything else.

At some point we had a finished product that we began to sell. We took the product to CES (the Consumer Electronics Show) and other home shows and we got a lot of positive feedback, but in order to sell the product in any of the big box stores we knew that we would also have to pass the rigorous testing at Underwriters Laboratories (UL). We met with a representative, but found him extremely restrictive and tangential. It was very frustrating to work with him because he was billing us for every hour we talked to him and for every question we asked he would go off on a lengthy, overly detailed discussion. He wouldn't even give coherent answers to the questions. When we got the quote for the work that UL said they would have to do we found it to be exorbitant. It felt like dealing with those lawyers that charge you for every phone call. Consequently, we went looking for some other company that could perform the same tests as UL only cheaper.

Eventually we did find another company, which will go unnamed here for reasons that will soon become clear. We then realized that there are these entities called Nationally Recognized Testing Laboratories (NRTLs) that are authorized to perform the same tests as UL to mark your product as safe. The lab that we chose was cheaper, more communicative and, most importantly, they could fit us into a tight schedule so that we could sell our product before the Christmas season. To our delight, after we submitted our product we were told that it passed all tests. They gave us the go ahead for our product, and we were off to the marketplace.

As we continued to go to trade shows, we would meet vendors that were very interested in our product. At one point we hit the jackpot. We met a representative from a company, whose name I will also omit, that was set to give us access to 3,000 big box stores. We were all ready to go. We had the molds. We had the fabrication house all set up. We were prepared for assembly, and we had the packaging ready. The company that we met with was even willing to fund the inventory and manage every aspect of the relationship with the big box stores. We could see that in a very short time we would be set for life.

There was just one little wrinkle. The company didn't trust the labs that had done the testing for us. They really wanted a UL certification on the product. Now that we were so close, the company even decided that they would pay for the UL testing for us, just to move the deal along. *Great,* we thought. *We're almost there. All we have to do is get that UL sticker, and we're home free.* We confidently submitted the product to UL. The other lab was supposed to have already performed the same exact standardized tests. This was just supposed to be a formality, but to our shock, our product didn't pass the UL testing. There was a problem with the electronics that meant the product was getting too hot. *How could that be?* We wondered. *They're the exact same tests.* Nevertheless, the product didn't pass.

We spent a bunch of money redesigning the electronics, but in the end we couldn't get the product to charge two tablets simultaneously. At some point we scoped the product down. The company that we had met with that was going to give us access to the big box stores decided to work with some vendors in China to produce a similar product. Somehow the multibillion dollar company wasn't afraid of our puny little patent in the slightest. They certainly wouldn't have dealt with us in the first place had we not had a patent, but when we failed to solve the electronics problem it wasn't worth it to them to continue working with us even with the patent.

The experience was very costly. I personally lost money, but fortunately it never hurt me too much since I had a lot of other sources of revenue. One of the partners wasn't so lucky. He lost his house and car because he had been putting in so much. We had been so close to really making a lot of money that he hadn't been doing anything else.

The experience data that I gained was very valuable. It gave me a far better feel for how everything works and the confidence that comes from familiarity with a situation. The fact is we were really close to making millions with our little product. If we had stayed with UL in the first place, they would have found the problem with our product much earlier and we probably would have been able to solve the problem in time. The experience I gained from the process goes even further. I know what happens when you have a booth at the CES trade show. I now know way more about what happens when you're trying to present something to the big box stores. I know a ton more about dealing with suppliers, UL testing, and electrical engineering, and I developed contacts with some amazing people.

MONEY AND BUYERS

Creative people sometimes throw caution to the wind. Some of us are adrenaline junkies. We throw ourselves into a new project with a beautiful vision and high hopes. In the process we sometimes underestimate what it will take to get a project done. We underestimate how long it will take. Perhaps this is the opiate we need in order to undertake a project that may be very large and difficult. Perhaps if we really knew up front how much effort we would end up putting into a thing, we would never do it. But in the end, underestimating can be deadly, particularly when it comes to funds.

A recent client of ours had an idea for a small electromechanical device. He was given an estimate for all the phases of the project, and he immediately up and signed. We all began working on the project enthusiastically, but as it progressed, the client had to stop it in its tracks. He found that there was no way to get the money required to move forward. It's always sad to see a promising idea fail due to a lack of funding. Don't let it happen to you.

FUND YOUR IDEA

One of the surest roads to abject failure is underfunding. Without the right amount of money you will have to compromise quality, or functionality, or safety, or proceed so slowly that by the time you come out with your product someone else will have done it before you, or worse, the whole need for your product will be gone entirely.

The first questions you need answers to are: How much will my product cost to produce? How much can I sell it for? What does the entire process look like, and what are the steps? How much does each step in the process cost? In order to answer these questions there are only a few ways to go. One way is by creating a flow chart of every necessary step: legal, industrial design, mechanical design, electrical design, manufacturing, packaging design, etc. Produce a Gantt chart[5] to illustrate your project schedule so that everything is clear with deadlines and action items.

---

[5] A **Gantt chart** is a type of bar **chart** that illustrates a project schedule. This **chart** lists the tasks to be performed on the vertical axis, and time intervals on the horizontal axis. The width of the horizontal bars in the graph shows the duration of each activity.

Create a business plan with as much detail as you can. There are endless quantities of resources out there to help you accomplish this, so go exploring. Find one that works for you. Once you have the business plan you are ready to go out and look for money. You need to figure out where the money is going to come from. If it's from savings or a trust fund, you're in great shape. If you need investors or loans, you have work to do. The most important question you need to ask when searching for money is: How much does the money cost? That is, what do you need to give back for it? Is it borrowed money so you can just pay back the principal plus interest, or is it investor money you pay for with ownership and control of the idea? Perhaps it's from an angel investor. Maybe it's money from friends and family.

No matter which way you go, you will have to present your idea very well. Hopefully you have a great prototype and a great way of sharing your idea. I hope you can be gregarious and persuasive, positive and strong, because you'll need to be. I hope you get the funding you need. I hope you use it to your great advantage and get your product done and on the market so we can all benefit from your idea.

TRADE SHOWS

Trade shows are a place where the magic happens for you and your product. You can potentially meet that company representative that will buy into your concept and help you get your product into 3,000 big box stores. On the other hand, if you're not careful, the trade show can be a huge waste of money and time. When trade shows are good, they bring together a collection of like-minded individuals who are all interested in the same industry. Many of your prospective customers will be right there in the same room along with people who will potentially be your competition. It's a huge wealth of knowledge all in one place. When trade shows are bad, they are poorly advertised and few people show up. Those that do will have little interest in your product, so it's very important that you're careful when navigating this complex space and all its aspects.

Should you choose to use this important selling technique, you'll begin your journey by finding the right trade show for you. Read up about them, ask around, read the reviews. Find out how many will attend and learn about the show's history. Trade shows vary widely in size, cost, frequency, and popularity. For example, the Consumer Electronics Show (CES) is held in Las Vegas at the beginning of every year and has been going on since 1967. As the name implies, it showcases the latest in electronic products. This includes just about everything you can think of: toys, automotive related products, home theater devices, and everything in between. The website reports that the latest show (as of this writing) was attended by as many as 182,000 folks. The latest products from companies like Samsung, Nokia, Bose, and Sonos were all showcased at the event. There were even small inventors showcasing far less sophisticated products. I recall seeing a booth that was selling a phone charger that you place your phone into before manually spinning the phone to generate a charge. I wasn't particularly impressed by the product, but they had a booth and were able to get valuable feedback on their invention.

Finding a trade show can be a little daunting since there are so many to choose from, but it's worth it to do your research. There are myriad home shows and manufacturing trade shows; it seems like any product that you can name has some sort of place at a show or expo somewhere. A quick Internet search will bring up the Toy Fair for toys and trade shows for growers, skateboarders, recreational vehicle enthusiasts, and many others.

When you decide on a trade show, you may choose to do a cost-benefit analysis. The costs can be high. You may pay $10,000 just for a ten-foot by five-foot booth, with another $5,000 for a pop-up that goes behind you. Hotel and travel for at least two people can cost you $5,000 more while shipping all of the materials and displays might pile on another grand or two. When you're done with the trade show, you may be able to purchase the attendees list, but it may also come at a very high price. All things considered, you can easily spend upwards of $15,000 for the entire experience. Make sure that it's worth it.

It's important to carefully consider how you set up your booth based on what you need. If you have people in your booth using the actual product, for example if you innovated a new type of VR headset, you'll need multiple people in the booth helping people use the device as well as others to gather the customer information. The booths that you can purchase usually come in a variety of sizes. The larger ones may have $100,000 displays in them and many folks hovering around to answer your every question. If you are a first-time inventor this is probably beyond what you need. If you rent one of the smallest booths and you sign up for the booth early, you can usually select the location. If possible, avoid the corners in the back. Try to get a booth right in the thick of things.

There are a number of other little tricks that you may be able to use in your booth. When people come to visit you, it's extremely important that you make efficient use of the time that you spend with them. Most trade shows make all kinds of helpful services available. One important service is the contact collection device. It's a device that you can point at the badge of any show attendee to send their information right into a database that is delivered to you as soon as the trade show ends. This is great because you don't have to collect a plethora of business cards to later manually type into some database. Also, the devices allow you to take notes that will help later such as, "This is the guy who said he had the store in NJ and wanted purchasing information."

Another trick that I've seen is the use of an entertainment activity. I've seen folks use dart boards, roulette wheels, miniature golf courses, and the like. My experience with these is that they tend to generate buzz, but it can then be difficult to convert the conversation about the game into a conversation about the product. It can be distracting and take a lot of valuable time to set up the game and help people to play it. If you use this trick, make sure the activity is extremely simple and needs no instruction.

You don't always have to attend as an exhibitor, and prices vary depending on how you choose to attend. If you choose to be an exhibitor, note that there are different levels of exhibitor that you can attend as. Otherwise, you may want to be a contributor and give some sort of seminar, or simply walk the floor and view all the products in your product category just to get a feel for what else is out there and how it's being sold. All exploration aspects are valuable in one way or another.

When you attend a good trade show, you will probably see products that compete with yours. This is a great thing because you get to see firsthand what your competitors claim about their product. You may even get to see it in use and hold it in your hands. You can ask questions about sales, and how the product is doing in various markets. Folks at the booths are invariably friendly and informative and may even tell you what they will be coming out with in the future. You may even be able to make a contact that enables you to sell your product to your competitor in the future.

Folks in the booth may be on any level. They may be salespeople hired just for that show and have very little power or decision-making ability, or they may be the CEO or inventor of the product that is fully authorized to make a deal with you. There will also be buyers at the show who will be delighted to have met someone with something new and different that can help them differentiate their store. From our experience with various trade shows, we've seen that some of the larger, well-financed companies sometimes have a suite attached to their booth where they are capable of having high level meetings with prospective suppliers and inventors. Make sure you're prepared. When you have something to sell, one of the first things you will be asked is whether you have a patent for it.

Most trade shows have breakout sessions. If you're lucky, these can be a gold mine for you as they have been for me. The breakout session is usually a small advertised event where you get to give a presentation of some product or high-level concept. For organizers it's a great deal because it's free content. Remember, the folks who put on trade shows want their trade show to be valuable to exhibitors and attendees alike. If you can offer some information of great value or even something that will be entertaining, it makes the show that much better.

Every year I attend a trade show called PLM World. It's a trade show where folks from all over the world come to learn about a certain suite of high-end software products from a company called Siemens PLM. For the last 26 years, I've attended and given a speech about one higher level CAD technique or another. It's been a great way to get the user community of Siemens software to remember me and know about the product that I have to offer. One of the products that I've been very successful with over the years is a series of books that I've written about the use of various CAD packages.

The breakout sessions are usually about 45 minutes, and in order to have the maximum impact, presentations must be interesting, fast moving, valuable, and positive. If you can make the attendees laugh a little and leave enough space in the presentation for good questions and answers it will work well for you. If you can come out from behind the lectern and talk about real experiences that you've had using your product or what happens when real users get it in their hands, if you can be extremely honest and forthcoming, you and your product will be remembered.

At the CES show, they have televised and podcasted reviews. They have a cadre of folks who are trying to make the show pop on social media. It's a great idea to meet with the media folks and offer your story as something that they may want to use. You, as the inventor of a new product, may have a very compelling story to tell. There are well-known product reviewers who have blogs and a voracious appetite for new material. These folks can help you to get the word out and net you that one chance meeting with someone who just happens to be a buyer from a popular outlet. Sometimes that's all it takes.

At many trade shows, there are luncheons for the attendees and exhibitors. You can make some great contacts at the luncheon. Whenever I attend a good trade show, I take the luncheon very seriously. It's not about the meal, it's about the contacts that you may make. If I'm there with some colleagues I'll insist that we all sit at different tables so we can be exposed to more people. We purposely choose tables where there are a variety of folks gathered who are hopefully not all from the same company. We look for tables where the people look like actual decision makers, and we always sit at a table where there are people already seated.

A few years ago I attended a trade show where I had a booth for three days straight and gave a speech. I was tired. I had been talking nonstop and working at night to perfect my speech. When the show was coming to a close, I was just about done. There was a luncheon. Usually I attend the luncheons with the idea that I might very well meet someone who can buy my services and books, someone I may be able to help. But I had left that on the proverbial playing field by then and I was beat. I purposely selected a table with as few people as possible and sat on the opposite end from the three people who were already there. The waiter brought me a meal, and I was glad to just sit and eat. Then I overheard the conversation of the three other people at the table. They were talking about problems that I was very familiar with and problems that I knew I could help them solve.

In an instant all the fatigue left my body and I perked up, shifting gears into a solving-problems mode. I joined their conversation, and we talked about solutions. One of the folks at the table said that he really liked what I had to say. On the spot he made the decision to do business with my company and said that there would be just one provision: It had to be me who came and performed the service, not one of my employees. It turned out that the man I had been talking to was a decision maker and the representative of a large Fortune 500 company. That little conversation led to a great friendship and a series of extremely lucrative contracts. So as you can see, the trade show luncheon can sometimes be absolutely transformational.

MORE EXPERIENCE DATA ON TRADE SHOWS

A few years ago I got together with some folks to come up with a new device. It was a simple product that we thought people would want; I've talked about it before. It was a product that made sense and was a reaction to a relatively new behavior: that people needed to charge more devices through a USB port than ever before. The device was a wall mounted USB phone or tablet charger. I personally did most of the mechanical engineering, and we had other folks do the electrical engineering and other tasks. We applied for a patent and after a time it was awarded, and we started a new LLC company called Currentwerks. Eventually we were ready to sell two new products. We named them the Duo and the Quattro. The idea was to pack enough technology into the cases that users could simultaneously charge a tablet and a phone at the full charging rate. This turned out to be quite the challenge due to temperature requirements as we found out about later.

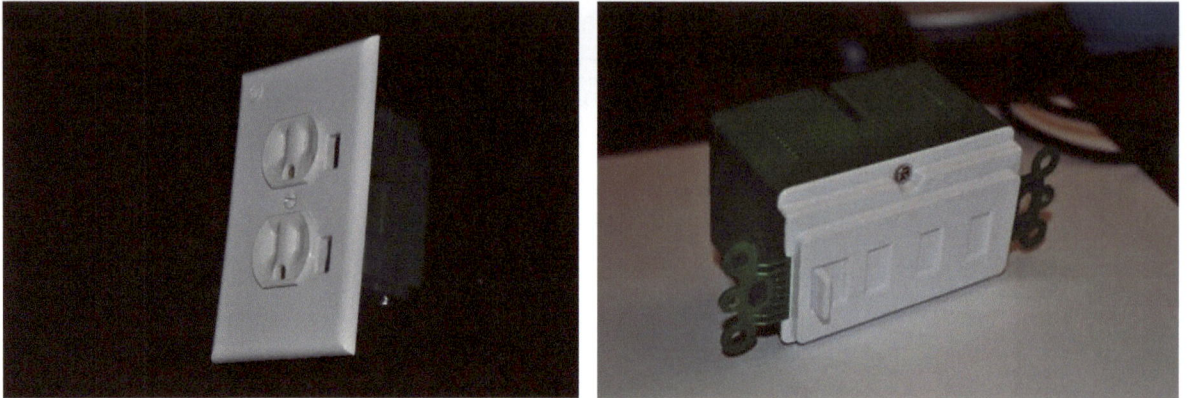

FIGURE 90

We made a display case, handcrafted out of maple with a glass inlay and a top where the product was mounted that lit up the device nicely and enabled prospective customers to handle it. The product literature was in little cubbies underneath. We attended the CES show, and we actually sold product off the showroom floor. This was probably more trouble than it was worth, but at least it created some buzz and informed potential buyers that we had a viable product.

FIGURE 91

We actually met an executive of a branding company who eventually offered us the ability to sell the product in 3,000 stores. We had to scramble to create packaging and get a number of other tasks done. At some point, we decided to create a more streamlined display that we could more easily take to some of the smaller home shows, and we came up with two desktop units that the product was mounted into. We met other folks at the home show that ordered large numbers of our product to sell in an online store.

FIGURE 92

At first, we were assembling our product at my home design office. After a time, we employed a contract manufacture to do the assembly. Purchasing the components and making deals with suppliers and all the other business functions took far more time and effort than the engineering. On the next page we have an example of a press release that helped in getting our information out there to distributers. When your product is ready to go, consider issuing your own press release and getting it to trade organizations, magazines, and websites that focus on your type of product.

CURRENTWERKS INTRODUCES ENERGY-SAVING USB WALL OUTLETS

Duo and Quattro Outlets Charge Portable Electronics Fast and Shut Down Power When Complete, Eliminating Vampire Power and Saving Energy Cost

Los Gatos, Calif., December 12, 2011—CurrentWerks, a leading innovator in power storage and charging devices, today introduced two powerful USB wall outlets, the Duo and Quattro. The company will exhibit these in booth #74010 in the Eureka Park TechZone for new and innovative products at the 2012 International CES, January 10–13 in Las Vegas.

The Duo adds two powerful USB ports with 16W of output to a standard 110V wall outlet. The Duo supports both 15A and 20A wall receptacles, making this a truly universal product.

The Quattro is a similarly unique product, with 4 USB ports that have a combined 22W of output that fit in a standard wall gang box. With its patented door mechanism, the Quattro completely eliminates vampire power saving enough energy to pay for itself within 1 year.

Both of these products are powerful enough to quickly charge even the most power-hungry devices, such as the iPad, and eliminate the need for wasteful and bulky AC adapters. The Duo and Quattro are both available for pre-order now and will be shipped in early January.

"These products are the latest innovations in green electricity," said Patrick Manning, CEO, CurrentWerks. "They are very easy to install and can replace existing outlets. Additionally, we have priced them very competitively making them the perfect solution for sustainable building in both home and commercial applications."

About CurrentWerks

Based in Los Gatos, Calif., CurrentWerks LLC is dedicated to providing leading innovations in power storage and charging devices for portable consumer electronics.

FINAL WORD

Congratulations once again! You've reached the end of this book! However, as you surely know by now, your journey as an inventor has only just begun. Now is the time for you to take action. Now is the time for you to actually go out there and work through everything you need to and bring your idea to market. Remember, there isn't much practical difference between someone who never has a good idea and someone who just doesn't act on it, so act! Have the confidence to make your invention a reality. Push yourself to have the drive to make it all come true. Get it out of your head, onto paper, and then out into the world. The world needs creative people like you who actively seek to improve it, even in the smallest of ways. So now, armed with the knowledge from this book in your hands, we beseech you: Go forth and invent!

BIBLIOGRAPHY

The Wisdom of Mentors

Note that all quotations are dredged from the fog of memory and, while truthful in spirit, may not be verbatim.

Anatomy of a Good Idea

"A Brief History of the Steel Pan." *BBC News*, BBC, 24 July 2012, www.bbc.com/news/magazine-18903131.

"General Information Concerning Patents." *United States Patent and Trademark Office*, United States Patent and Trademark Office, 14 Feb. 2018, 2:34 PM EST, www.uspto.gov/patents-getting-started/general-information-concerning-patents#heading-4.

"How Steve Jobs' Love of Simplicity Fueled a Design Revolution." *Smithsonian.com*, Smithsonian Institution, 1 Sept. 2012, www.smithsonianmag.com/arts-culture/how-steve-jobs-love-of-simplicity-fueled-a-design-revolution-23868877/.

Kersting, Karen. "What Exactly Is Creativity?" *Monitor on Psychology*, American Psychological Association, Nov. 2003, www.apa.org/monitor/nov03/creativity.aspx.

Lubow, Arthur. "Edvard Munch: Beyond The Scream." *Smithsonian.com*, Smithsonian Institution, 1 Mar. 2006, www.smithsonianmag.com/arts-culture/edvard-munch-beyond-the-scream111810150/.

Martin, Gary. "'Genius Is One Percent Inspiration, Ninety-Nine Percent Perspiration'—The Meaning and Origin of This Phrase." *Phrasefinder*, www.phrases.org.uk/meanings/genius-is-one-percent-perspiration-ninety-nine-percent-perspiration.html.

Mulkerrins, Jane. "Aaron Sorkin: Author of The Newsroom on His New Steve Jobs Movie." *The Independent*, Independent Digital News and Media, 17 Nov. 2014, www.independent.co.uk/arts-entertainment/tv/features/aaron-sorkin-interview-writer-of-the-newsroom-on-his-battle-with-drugs-and-being-a-perfectionist-9862342.html.

Pearce, Jeremy. "Arthur Galston, Agent Orange Researcher, Is Dead at 88." *The New York Times*, The New York Times, 23 June 2008, www.nytimes.com/2008/06/23/us/23galston.html.

Rosenthal, Arthur. "The History of Calculus." *The American Mathematical Monthly*, vol. 58, no. 2, 1951, pp. 75–86., doi:10.1080/00029890.1951.11999628.

Runco, Mark A. *Creativity: Theories and Themes: Research, Development, and Practice*. 2nd ed., Academic Press, 2014.

Runco, Mark A. "Personality and Motivation." *Creativity: Theories and Themes: Research, Development, and Practice*, 2nd ed., Academic Press, 2014, pp. 265–302.

Shapira, Oren. "An Easy Way to Increase Creativity." *Scientific American*, Scientific American, 21 July 2009, www.scientificamerican.com/article/an-easy-way-to-increase-c/.

Strada, Gino. "The Horror of Landmines." *PBS*, Public Broadcasting Service, 18 Jan. 2002, www.pbs.org/pov/afghanistanyear1380/the-horror-of-landmines/.

Suddath, Claire. "A Brief History of: Velcro." *Time*, Time Inc., 15 June 2010, content.time.com/time/nation/article/0,8599,1996883,00.html.

"The Creative Power of 'Outsiders.'" *Association for Psychological Science*, Association for Psychological Science, 5 Apr. 2016, www.psychologicalscience.org/news/minds-business/the-creative-power-of-outsiders.html.

"Water Roller." *Hippo Roller*, Imvubu Projects (Pty) Ltd., www.hipporoller.org/water-roller/.

Special thanks to Hippo **Roller.org** for their permission to use their images.

Product Design

Apple, iPod, iPad, and iPod nano are registered trademarks of Apple Inc.

Devlin, Keith. "Unravelling the Myth." *The Guardian*, Guardian News and Media, 12 Sept. 2001, www.theguardian.com/science/2001/sep/13/physicalsciences.highereducation.

Dougherty, Philip H. "Advertising; The Birth of Apple's Ad Insert." *The New York Times*, The New York Times, 1 Mar. 1984, www.nytimes.com/1984/03/01/business/advertising-the-birth-of-apple-s-ad-insert.html.

Garson. "Sometimes a Cigar Is Just a Cigar." *Quote Investigator*, Quote Investigator, quoteinvestigator.com/2011/08/12/just-a-cigar/.

"How Steve Jobs' Love of Simplicity Fueled a Design Revolution." *Smithsonian.com*, Smithsonian Institution, 1 Sept. 2012, www.smithsonianmag.com/arts-culture/how-steve-jobs-love-of-simplicity-fueled-a-design-revolution-23868877/.

Hand Sketching

Eissen, Koos, and Roselien Steur. *Sketching: Drawing Techniques for Product Designers*. BIS Publishers, 2015.

Protecting Your Idea

All prices and fee estimates are taken of the USPTO website, and are accurate at the time of writing.

Bilski v. Kappos, 130 S. Ct. 3218, 561 U.S. 593, 177 L. Ed. 2d 792 (2010).

Working with Others

"3M's 15% Culture." *3M Careers*, 3M, www.3m.com/3M/en_US/careers-us/culture/15-percent-culture/.

iPad is a registered trademark of Apple Inc.

McCullough, David. *Truman*. Simon & Schuster, 2003.

Images

"1372470." *Pxhere*, 06 Apr. 2017, pxhere.com/en/photo/1372470.

"551304." *Pxhere*, 26 Jan. 2017, pxhere.com/en/photo/551304.

"628721" *Pxhere,* 06 Feb. 2017, pxhere.com/en/photo/628721.

Ahanix1989. "Ricerburner-Neon." *Wikimedia Commons*, 10 Feb. 2006, commons.wikimedia.org/wiki/File:Ricerburner-neon.jpg.

Artisan Entertainment ; director, George P. Cosmatos. Rambo : First Blood Part II. Santa Monica, Calif. :Artisan Home Entertainment, 2002. Print.

Aust, Herbert. "Ipod Nano Apple." *Pixabay.* 20 Oct. 2016, pixabay.com/en/ipod-ipod-nano-apple-nano-1752964/.

Beall, C. C. "7th War Loan. Now All Together." 1945, U.S. Government Printing Office.

Beem, Alf van. "Art and Tech VERA concept car at the Musee Automobile de Vendee." *Wikimedia Commons,* 14 Sept. 2013, commons.wikimedia.org/wiki/File:Art_%26_Tech_VERA_concept_car_at_the_Musée_Automobile_de_Vendée_pic-05.JPG.

Bonzanigo, G. & Pergolesi, M. "Console Table." 1782–1792. Metropolitan Museum of Art.

Danjestorres. "Slippers White Glamour." *Pixabay,* 7 July 2017, pixabay.com/en/slippers-white-glamour-women-2480251/.

Derewecki, Adam. "Neuschwanstein, castle." *Pixabay.* 16 Nov. 2014, pixabay.com/en/neuschwanstein-castle-germany-532850/.

Deutsch. "Chrysler-building-new-york-1167187." *Pixabay,* 1 Feb. 2016, pixabay.com/en/chrysler-building-new-york-1167187/.

Diniz, Luis. "Venis Statue Louver." *Pixabay.* 22 Dec. 2014, pixabay.com/en/venus-statue-louvre-paris-575092/.

Gavilla, Guillermo. "David Michelangelo Statue." *Pixabay.* 23 Mar. 2016, pixabay.com/en/david-michelangelo-statue-italy-1270344/.

Hippo Roller. "Women Pushing Rollers." *Flickr,* 26 Mar. 2008. www.flickr.com/photos/64491163@N06/15011729888.

Hudson, Dawn. "We Can Do It." *PublicDomainPictures.net,* www.publicdomainpictures.net/en/view-image.php?image=76000&picture=we-can-do-it-poster.

Kratochvil, Petr. "Hibiscus Flower." *PublicDomainPictures.net,* www.publicdomainpictures.net/en/view-image.php?image=4305&picture=hibiscus-flower.

Lesch, Susan G. "Biosphere in Montreal." *Wikimedia Commons,* 8 Jan. 2007, commons.wikimedia.org/wiki/File:Biosphere_in_Montreal.jpg.

Lowe, Jet. "LOC Brooklyn Bridge and East River 7." *Wikimedia Commons,* 23 May 2009, commons.wikimedia.org/wiki/File:LOC_Brooklyn_Bridge_and_East_River_7.png.

M 93. "Cadillac CTS 2.0 Turbo Luxury (III)." *Wikimedia Commons,* 5 Sept 2015, commons.wikimedia.org/wiki/File:Cadillac_CTS_2.0_Turbo_Luxury_(III)_-_Frontansicht,_5._September_2015,_Düsseldorf.jpg.

Manfredsson, Cecilia. "Athens the Pantheon Greek." *Pixabay,* 3 July 2013, pixabay.com/en/athens-the-parthenon-greek-greece-142761/.

Misra, Rajesh. "Hall with Hundred Pillars." *PublicDomainPictures.net,* ww.publicdomainpictures.net/en/view-image.php?image=147952&picture=hall-with-hundred-pillars.

Monet, Claude. *Apples and Grapes.* 1880.

Monet, Claude. *Bed of Chrysanthemums.* 1897.

Onuka, Masami. "Victim of Atomic Bomb 002." *Wikimedia Commons*, 10 Sept. 2005, commons.wikimedia.org/wiki/File:Victim_of_Atomic_Bomb_002.jpg.

Powell, Mike. "Erika, Beach 10." *Flicker,* 6 Oct. 2008, www.flickr.com/photos/22170282@N05/2920127540.

Prayitno. "Hope Diamond." *Flicker,* 17 Sept. 2012, www.flickr.com/photos/prayitnophotography/8123979570.

Puiforcat, Jean. "Puiforcat, Teeservice Art Deco." *Wikimedia Commons*, 1 Jan. 1972, commons.wikimedia.org/wiki/File:Puiforcat,_Teeservice_Art_Deco.jpg.

Rbramosjr. "Great Wall of china-mutianyu 1." *Wikimedia Commons*, 12 July 2006, commons.wikimedia.org/wiki/File:Great_wall_of_china-mutianyu_1.jpg.

Renoir, Pierre-Auguste. *Bather on a Rock.* 1892. Sammlung Durand-Ruel, Paris (Private collection).

Rubens, Peter. *The Three Graces.* 1630–1635. Museo del Prado, Madrid.

Tentis, Dana. "Newborn, Feet, Hands." *Pixabay.* 4 Aug. 2017, pixabay.com/en/newborn-feet-hands-cute-adorable-2653072/.

Thefss. "Cartier Panther Head Diamond." 13 Jan. 2016, pixabay.com/en/cartier-panther-head-diamond-1137400/.

Tiluria. "Owl Wood Carving." *Pixabay.* 15 Oct. 2017, pixabay.com/en/owl-wood-carving-figure-artwork-2855195/.

Tunechick85. "Pregnancy Belly Baby Expect." *Pixabay.* 3 Apr. 2017, pixabay.com/en/pregnancy-belly-baby-expecting-2196931/.

Wicker Paradise. "Wicker Chair." *Flicker,* 30 Apr. 2013, www.flickr.com/photos/wicker-furniture/8695611158/.

Wicker Paradise. "La Luna Rattan Chair." *Flickr,* 10 Feb. 2013, www.flickr.com/photos/wicker-furniture/8461753721/.

Wicker Paradise. "Real Mahogany Wood Iphone Skin Sticker." *Flickr,* 18 Dec. 2013, www.flickr.com/photos/wicker-furniture/ 11439328536/.

Williams, Vaughan. "Abencerrajes." *Flickr,* 14 Sept. 2004, de.wikipedia.org/wiki/Datei:Abencerrajes.jpg.